Dedicated to my parents.
Thank you for your endless love and support.

Contents

Preface xiii

ASTRONOMY 1

Planets 3
Exoplanets 8
The Moon 10
Moons—General 13
Meteors, Comets, and Asteroids 15
The Sun 18
Stars 21
The Milky Way Galaxy 25
Galaxies 26
Solar Storms 28
Supernovas 29
Neutron Stars 31
Gamma-Ray Bursts 35
Cosmic Rays 36
Neutrinos 38
Telescopes 39
Miscellaneous Astronomy 41

BIOLOGY....45
Plants....47
Plant Perception....50
Flowers....53
Trees....54
Fungi....57
Agriculture....59
Viruses....62
Bacteria....63
Microorganisms....67
Life....69
Chirality....71
Genealogy....73
Human Genetics....75
Genetics—General....78
Brain Organoids....81
Miscellaneous Biology....83

CHEMISTRY....87
Chemicals....89
Elements....91
Helium....93
Metals....95
Gold....97
Diamonds....98
Water....99
Food Chemistry....102
Pharmacology....103
Disappearing Polymorphs....106

The Haber-Bosch Process 108
People and Chemistry 111
Miscellaneous Chemistry 112

COSMOLOGY 117
The Big Bang 119
Dark Energy 122
Dark Matter 123
Types of Black Holes 124
Black Hole Traits 127
Approaching a Black Hole 130
The Size of the Universe 131
The Multiverse 134
The Simulation Hypothesis 136
Boltzmann Brains 139
False Vacuum Decay 140
The Fate of the Universe 141
Miscellaneous Cosmology 144

EARTH SCIENCE 147
Earth's Magnetic Field 149
Earth's Rotation 151
Earth's Interior 153
The Atmosphere 154
Lightning 157
The Water Cycle 160
The Ocean 162
Volcanoes 164
Earthquakes 165

Natural Disasters—Other 167
Geology 168
Fossil Fuels 172
Microplastics 174
Anthropology 176
Earth and Space 178
The Chicxulub Impactor 179
Miscellaneous Earth Science 183

HUMAN BODY 187

Perception Limits 189
The Mouth 191
The Nose 194
Ears 196
Eyes 198
Visual Phenomena 200
Blindness 206
Gastrointestinal 208
Respiratory 209
Cardiovascular 210
The Brain 211
Neurological Phenomena 215
Reflexes 219
Internal Organs 220
Skin 221
Hands 223
Bones 224
Muscles 226
Fat 229

Reproduction . . . 230
Pregnancy . . . 233
Babies . . . 237
Sleep . . . 240
Cancer . . . 241
Unusual Conditions . . . 242
Human Evolution . . . 245
Miscellaneous Human Body . . . 248

PHYSICS . . . 257

Physics Phenomena . . . 259
Acoustics . . . 260
Optics . . . 263
The Electromagnetic Spectrum . . . 266
Thermodynamics . . . 267
Nuclear . . . 269
Motion . . . 271
Time . . . 274
Planck Values . . . 275
Subatomic Particles . . . 277
Quantum Mechanics . . . 281
Miscellaneous Physics . . . 285

TECHNOLOGY . . . 289

Computing . . . 291
The Future of Computing . . . 293
Artificial Intelligence . . . 295
Phones . . . 296
Satellites . . . 298

Space and Technology 299
Space Transportation 302
Transportation.. 307
Construction 309
Military... 311
Surveillance 313
Audio 315
Microwaves 316
Nanotechnology 317
Environmental Technology........................... 319
Medical Technology...................................... 322
Information 325
Laboratory... 326
Historical Inventions 328
Modern Inventions.. 330
Miscellaneous Technology 332

ZOOLOGY... 337
Fish 339
Sharks.. 343
Whales 344
Octopuses.. 347
Shrimp 349
Aquatic Animals.. 350
Snakes.. 355
Reptiles.. 356
Birds 358
Bees.. 363
Ants.. 365

Insects....367
Parasites....373
Worms....375
Spiders....377
Rodents....378
Dogs....380
Cats....383
Elephants....385
Primates....387
Extinct Animals....390
Dinosaurs....393
Mammalian Traits....395
General Animal Traits....397
Miscellaneous Zoology....400

Preface

For as long as I can remember, I have been spellbound by science.

As a young child, I continuously bombarded my parents with the standard "*Why?*" questions. These included "*Why is the sky blue?*", "*Why did the dinosaurs go extinct?*" and "*Why do plants need sunlight?*" My parents would patiently answer each question, to which I would inevitably respond, "*Well, then why is that the case?*" To avoid a potentially infinite conversation, they'd eventually resort to the timeless response: "*That's just the way it is*".

As I grew older, my curiosity turned into action. I became obsessed with conducting my own experiments. Some were from kits—magnetism, chemistry, and crystal-growing—while others were homemade attempts to understand the world. My friends and family developed an instant scepticism whenever I asked them to partake in anything, always wary that they might unwillingly be part of some experiment.

As an adult, I pursued astrophysics at university and dabbled in some medical writing. However, my childlike curiosity never faded. Every day, I find myself asking dozens of scientific questions, many echoing the ones my younger self posed to my parents. The difference is that now I can seek out the answers myself, sparing my parents an endless interrogation.

These incessant "*Why?*" questions inspired me to write this book: the world's first encyclopedia of science oddities. My goal was to produce a collection of scientific oddities, each of which evokes that delightful 'wow' moment. They include a mixture of the counterintuitive, the surprising, the awe-inspiring, and the downright strange. This encyclopedia features nine chapters—Astronomy, Biology, Chemistry, Cosmology, Earth Science, Human Body, Physics, Technology, and Zoology—each with its own assortment of subtopics.

Writing this encyclopedia has been a labour of love—a joy, yet an immense amount of work (I don't think people would believe me if I told them the sheer amount of time I devoted!). This has involved reading hundreds of scientific studies and researching countless different sources. I've gone to great lengths to ensure that every oddity in this encyclopedia is both true and accurate. I've also strived to explain each concept as clearly as possible, avoiding overly technical jargon to make the content accessible to all.

This encyclopedia is for anyone who still has that childlike wonder and curiosity. It's for those who never grew out of asking "*Why?*"

ASTRONOMY

Planets

Contrary to popular belief, the closest planet to Earth on average is not Venus or Mars—it's actually Mercury. Earth spends 46% of its time closest to Mercury, 36% to Venus, and 18% to Mars.

★ ★

On Venus, you could theoretically watch the sunset forever by walking towards it at a steady pace. This is possible because the planet rotates extremely slowly, taking 243 Earth days to complete a single rotation.

★ ★

Venus takes longer to rotate once on its axis than to complete a single orbit around the Sun. Consequently, a day on Venus is longer than a year.

★ ★

If you stood on the surface of Venus, you would be instantly crushed by the overwhelming weight of the atmosphere—it would be the equivalent of having a small car pressing down on every square inch of your body.

★ ★

The Martian volcano Olympus Mons has such a gradual average slope—just five degrees—and such an immense height—22 kilometres—that somebody standing at its base wouldn't be able to see its summit because it lies beyond the horizon.

★ ★

If you stood on the equator of Mars at noon, it would feel like summer at your feet and winter at your head. This is because the Martian atmosphere is so thin that heat easily escapes the planet, resulting in a sharp temperature gradient between the warm surface and the cooler air just above it.

★ ★

If you poured a bottle of water on Mars, the water would boil and freeze simultaneously.

★ ★

In 2024, scientists discovered evidence of a vast reservoir of liquid water hidden beneath the surface of Mars. Located between 11.5 and 20 kilometres deep, the water is so abundant that it could cover the entire surface of the planet to a depth of 1.6 kilometres.

★ ★

The Great Red Spot on Jupiter is the largest storm in the Solar System. It has a diameter of 16,350 kilometres, making it 1.3 times the size of Earth. Although the storm has been raging for hundreds of years, it has recently been shrinking. In fact, scientists predict that it could disappear entirely within the next few decades.

★ ★

The exoplanet J1407b boasts a ring system 200 times larger than Saturn's. If Saturn had rings of this magnitude, they would be clearly visible from Earth with the naked eye, appearing even larger than the full Moon.

★ ★

In 2025, the James Webb Space Telescope detected preliminary evidence of high levels of dimethyl sulphide in the atmosphere of the exoplanet K2-18b. On Earth, this gas is produced solely by marine phytoplankton and bacteria, raising the possibility that life may also be producing it on the exoplanet. If life is confirmed on K2-18b, it would have profound implications. Not only would it rank as arguably the most significant discovery in the history of science, but the close proximity of the planet—just 143 light-years from Earth—would indicate that life is likely abundant throughout the Universe.

★ ★

A doughnut-shaped planet is theoretically possible.

★ ★

A rogue planet is a planet that drifts through space without being bound to a parent star. It's estimated there are more rogue planets in the Universe than traditional star-orbiting planets. This means, within the Milky Way alone, there are likely trillions of these rogue planets freely floating through space.

★ ★

There are more planets in the observable Universe than seconds that have passed since the Big Bang.

The Moon

The Selenean summit is the highest point on the Moon's surface. With an elevation of 10,786 metres above the lunar mean, it's nearly two kilometres higher than the summit of Mount Everest.

★ ★

The coldest known place in the Solar System is Hermite Crater, located near the north pole of the Moon. A portion of this crater that never receives sunlight experiences temperatures as low as -249 degrees Celsius. In comparison, on Pluto's surface, which is known for its cold conditions, temperatures only drop to -229 degrees Celsius.

★ ★

The Moon is covered in a layer of toxic dust that is as fine as powder, yet as sharp as glass. This abrasive substance is so harsh that it cut through the spacesuits and boots worn by the Apollo astronauts.

★ ★

Have you ever gazed up at the Moon and wondered what those large, dark patches on its surface are? These are actually vast seas made of ancient, dried lava.

★ ★

Across billions of years, flowing lava on the lunar surface carved out hollow, underground chambers known as lava tubes. These ancient formations, which still exist today, are the perfect location for future research stations or lunar settlements. They offer a stable visible due to the intense friction they generate as they travel at high speeds through Earth's atmosphere. This friction heats the surrounding air, causing it to glow and creating the bright streak we see in the night sky.

★ ★

You shouldn't be able to hear a meteor until several minutes after sighting it, since sound travels about one million times slower than light. However, astronomers have recently discovered an exception to this rule: electrophonic meteors. These meteors emit radio waves that travel at the speed of light. While the radio waves themselves are inaudible, they can cause objects on Earth's surface to vibrate, creating sounds that you can hear. Thanks to this phenomenon, it's possible to hear the sound of a meteor at the exact moment you see it streaking across the sky.

★ ★

In 1969, the Murchison meteorite landed in Australia, carrying seven-billion-year-old silicon carbide particles. These particles are the oldest material ever discovered on Earth, predating the formation of the Solar System by 2.5 billion years.

★ ★

Micrometeorites are pieces of space rock, smaller than a grain of sand, that have landed on Earth's surface. These particles often date back to the birth of the Solar System, yet they are surprisingly common. In fact, an average of ten micrometeorites land on every square metre of Earth's surface each year. Despite their extraterrestrial origin and status as the most exotic small rocks in the Universe, micrometeorites can often be found in something as mundane as your roof's gutter.

The Sun

The Sun makes up 99.86% of the mass of the Solar System.

★ ★

When the Sun formed 4.6 billion years ago, it likely had a companion star. Over time, this second star was ejected from the Solar System, leaving the Sun as the single-star system we know today.

★ ★

Throughout the Sun's 4.6-billion-year lifetime, it has completed just 20 orbits around the Milky Way.

★ ★

The Sun doesn't just orbit the centre of the Milky Way—it also moves above and below the galactic plane. At present, it's located 55 light-years above this plane.

★ ★

If the Sun were scaled down to the size of a grain of sand, the nearest star, Proxima Centauri, would be 14 kilometres away.

★ ★

The Sun is actually white. It appears yellow to us because we view it through Earth's atmosphere. The astronauts aboard the International Space Station see the Sun as its true colour—white.

★ ★

The Sun experiences a phenomenon called differential rotation, where different latitudes of its surface rotate at different speeds. For example, at the Sun's equator, a single rotation takes 25 Earth days, whereas at its poles, a single rotation takes 36 Earth days. This difference in rotational speed contributes to the formation of sunspots, solar flares, and coronal mass ejections.

★ ★

There is water on the Sun. The dark patches on the Sun's surface, called sunspots, contain trace amounts of water vapour.

★ ★

The Sun contains roughly one atom of gold for every trillion atoms of hydrogen. While this might seem like an incredibly small proportion, the Sun's sheer mass means that it still holds two quadrillion tonnes of gold—enough to fill the Caspian Sea.

★ ★

Counterintuitively, the Sun's atmosphere is thousands of times hotter than its surface. While the surface of the Sun has a temperature of 5,500 degrees Celsius, the temperature soars to millions of degrees Celsius as you travel into the Sun's atmosphere.

★ ★

Every millisecond, the Sun emits more energy than humans have consumed throughout the entirety of civilisation.

★ ★

The Sun's core produces the same amount of power per cubic metre as a compost heap.

★ ★

A photon takes 100,000 years to travel from the Sun's core to its surface. It then takes just eight minutes and twenty seconds to travel the remaining distance to Earth.

★ ★

If sound could travel through space, we would hear the Sun roaring at approximately 100 decibels—a volume nearly as loud as a rock concert. This incessant barrage of noise would make it difficult to hold daytime conversations on Earth.

★ ★

The Sun and the Moon appear the same size in our sky—a coincidence not shared by any other known planet-moon combination. It occurs because the Sun is 400 times larger than the Moon and 400 times farther away.

★ ★

Earth is gravitationally attracted to where the Sun is now, not where it was eight minutes ago.

★ ★

The likelihood of another star ever colliding with the Sun is astronomically low. In fact, if four snails were randomly placed in the continental United States, there would be a greater chance of two of them accidentally bumping into each other than of another star colliding with the Sun.

Stars

Approximately 99.9% of matter in the visible Universe isn't solid, liquid, or gas—it's actually plasma. This is because stars, which are almost entirely composed of plasma, account for the overwhelming majority of observable matter.

★ ★

We live at the tail end of star formation in the Universe. In fact, around 95% of all the stars that will ever exist have already been born.

★ ★

During the cosmic noon—a period two to three billion years after the Big Bang—the star formation rate was ten times higher than it is today.

★ ★

The observable Universe contains up to one septillion stars—that's one followed by 24 zeroes. To put this immense number into perspective, if each star were represented by a grain of sand, they would cover the entire Earth's surface to a depth of your waist.

★ ★

In the time it takes you to blink, approximately 1,000 new stars are born in the observable Universe.

★ ★

Roughly one-third of all star systems in the Universe are binary or multiple-star systems, where two or more stars orbit each other. The remaining two-thirds are single-star systems, like our Solar System.

★ ★

Some astronomers believe that every star begins its life as part of a binary system. According to this theory, when a cloud of gas and dust collapses, it splits into two fragments, each forming a star. Although not all star pairs remain bound together—some drift apart over time—this theory might explain why binary star systems are so common throughout the Universe.

★ ★

Stars can exist inside other stars. For example, a supergiant star can have a neutron star at its core.

★ ★

Counterintuitively, larger stars have shorter lifespans than smaller stars. This is because the larger a star's mass, the faster it burns through its nuclear fuel.

★ ★

Low-mass red dwarf stars can burn continuously for ten trillion years. That's roughly 700 times the current age of the Universe.

★ ★

VY Canis Majoris is a truly colossal star located just 3,900 light-years away. To put its immensity into perspective, if Earth were scaled down to the size of a tennis ball, VY Canis Majoris would be as tall as Mount Everest. Given VY Canis Majoris's enormous size, if a jet circumnavigated it at 900 kilometres per hour, it would take 800 years to complete the journey.

★ ★

UY Scuti is one of the largest known stars in the Universe. It's so massive that you could fill it with five billion Suns or seven quadrillion Earths. If UY Scuti were placed at the centre of our Solar System, it would extend beyond the orbit of Jupiter. Even travelling at the speed of light, it would take seven hours to circumnavigate it.

★ ★

The first generation of stars to form in the Universe were significantly larger than present-day stars.

★ ★

R136a1 is one of the most luminous known stars in the Universe, with a luminosity 4.7 million times greater than the Sun. In just four seconds, it emits more energy than the Sun does in an entire year. If R136a1 were as close to Earth as Proxima Centauri, it would appear in our sky brighter than the full Moon.

★ ★

S4714 is the fastest-moving star ever observed. It orbits Sagittarius A, the supermassive black hole at the centre of the Milky Way, at an astonishing 8% of the speed of light.

★ ★

Around 70,000 years ago, our Solar System briefly contained three stars when a binary star system, known as Scholz's Star, passed through the Oort cloud.

★ ★

On average, a star passes through the Oort cloud once every 100,000 years.

★ ★

Przybylski's Star is a rapidly oscillating Ap star located 356 light-years from Earth. It is considered one of the most unusual stars in the Universe because it appears to contain a variety of rare and radioactive elements that are not found in any other known star. For instance, astronomers have detected promethium in Przybylski's Star, a radioactive element with a half-life of just 17.7 years. Since promethium decays so quickly, its presence suggests that some unknown process must be continuously replenishing it in Przybylski's Star. There is also some evidence that Przybylski's Star contains einsteinium, a synthetic element that's only ever been created by humans. In addition to these elements, Przybylski's Star might also harbour actinium, americium, berkelium, californium, curium, neptunium, plutonium, and protactinium. The reason why there are so many exotic elements in Przybylski's Star is a mystery.

The Milky Way Galaxy

If the Sun were scaled down to the size of a single white blood cell, the Milky Way would be as large as Russia.

★ ★

On average, only seven new stars are born in the Milky Way each year.

★ ★

The Milky Way is one of the more densely packed regions in the Universe. Yet even here, matter is so sparse that for every one part of matter, there are 24 sextillion parts of empty space.

★ ★

There is a 50% chance that the Milky Way and Andromeda galaxies will merge within the next ten billion years. While the Milky Way contains 400 billion stars and Andromeda one trillion stars, the vast distance between stars means that the likelihood of any two stars colliding during this merger is incredibly small—just 0.00000004%. To put this value into perspective, there's a greater chance of winning the Australian national lottery than of a single star collision occurring.

★ ★

Although the merger between the Milky Way and Andromeda galaxies may commence in the distant future, the process has technically already begun. This is because both galaxies are surrounded by their own vast galactic halo, composed of dust, gas, and stray stars. Astronomers have measured Andromeda's halo and discovered that it extends halfway to the Milky Way. Although the Milky Way's halo cannot be measured directly, it's believed to be a similar size. As a result, the two halos are likely already overlapping.

★ ★

It's a common myth that many of the stars we see in the sky are already dead. In reality, only about 4,000 stars are visible to the naked eye from Earth, and all of them lie within 4,000 light-years of us, in our local region of the Milky Way. Since most stars have lifespans ranging from millions to billions of years, it's highly unlikely that any stars you can see with your naked eye have already died.

Galaxies

Ten trillionths of your suntan comes from stars outside the Milky Way.

★ ★

When you hold a single grain of sand at arm's length, it covers 10,000 galaxies in the sky.

★ ★

The Canis Major Dwarf Galaxy is closer to us than the centre of our own Milky Way Galaxy.

★ ★

Despite being the closest known galaxy to the Milky Way, the Canis Major Dwarf Galaxy remained undiscovered until 2003 because it was obscured by dust within the Milky Way's galactic plane.

★ ★

The Andromeda Galaxy, located 2.5 million light-years away, is the farthest object visible with the naked eye. The photons hitting your eye when you gaze at Andromeda began their journey 2.5 million years ago. That's around the time the first species of the Homo genus began to evolve.

★ ★

The Dragonfly 44 galaxy is nearly invisible. Despite having the same mass as the Milky Way, it contains virtually no discernible stars. Astronomers believe that it's composed almost entirely of dark matter.

★ ★

The galaxy MoM-z14 is the most distant astronomical object ever detected. Located 33.8 billion light-years away, it lies nearly three-quarters of the distance to the edge of the observable Universe. The light we see from MoM-z14 began its journey towards us just 280 million years after the Big Bang.

★ ★

Galaxies appear smaller in the sky the farther they are away, as you would expect. However, beyond a certain distance, something counterintuitive happens: galaxies actually appear larger in the sky the farther they are away. This occurs because when we observe extremely distant galaxies, we are seeing the light they emitted billions of years ago, at a time when the Universe was much smaller and those galaxies were closer to us. Due to this effect, some of the most distant galaxies we've ever observed, like JADES-GS-z13-0, appear twice as large in the sky as similarly-sized galaxies that are only half the distance away.

★ ★

Galaxies are so sparsely filled with matter that if you were to travel across the diameter of one in a spacecraft, the probability of colliding with any object, planet-sized or larger, would be virtually zero. In fact, you would need to travel a distance equivalent to crossing the observable Universe 6,300 times before you would expect to have your first collision.

Solar Storms

The Carrington Event was a large solar storm that reached its peak on 1-2 September 1859. During this event, the Sun released a powerful coronal mass ejection—a burst of plasma and magnetic field—that collided with Earth's magnetosphere, triggering auroras around the world at unusually low latitudes, including in Mexico, Hawaii, and Cuba. These auroras produced enough light for those in the northeastern United States to read the newspaper during the middle of the night. They were so bright that they even woke coal miners in Colorado, who began making breakfast, mistakenly believing it was dawn.

★ ★

During the Carrington Event, some telegraph operators were able to maintain a conversation, despite having their batteries disconnected. This was made possible because the telegraph lines became powered by the intense geomagnetic activity of the solar storm.

★ ★

On 23 July 2012, the Sun unleashed a powerful coronal mass ejection that narrowly missed Earth by just nine days. Had it struck our planet, the consequences would have been catastrophic. It would have severely damaged power grids, satellites, and communication networks on a global scale. The financial impact of such an event is estimated to have reached trillions of dollars, potentially crippling the global economy. Overall, it may have taken up to a decade to recover from the damage.

★ ★

Every year, there is a 4% chance that Earth will be hit by a severe solar storm, and a 0.7% chance that it will experience one as intense as the Carrington Event.

Supernovas

A supernova is a large explosion that occurs when a massive star reaches the end of its life. This spectacular event is so bright that it can outshine an entire galaxy. In the Milky Way—with its 400 billion stars—a supernova explodes on average once every 50 years. However, across the entire observable Universe, a supernova explodes on average once every ten seconds.

★ ★

Astronomers can detect supernovas in our galaxy hours before their light reaches us. This is made possible because supernovas release a massive burst of neutrinos—particles that barely interact with matter—that escape the collapsing star almost instantly. In contrast, the photons from the explosion take much longer to emerge because they must push through the star's dense layers. As a result, the neutrinos reach Earth hours before the first visible light from the explosion arrives.

★ ★

If a supernova were to explode at the distance of the Sun, all life on Earth would perish before the light or blast wave reached us. This is because a supernova releases such an immense number of neutrinos—about one quadrillion times more than the Sun—that all life would be boiled from the inside out. While the probability of any single neutrino interacting with an atom in a person's body is incredibly low, the sheer number of neutrinos passing through a person would result in countless interactions.

★ ★

If a supernova exploded at the distance of the Sun, it would be one billion times brighter than a hydrogen bomb detonated against your eyeball.

★ ★

A supernova named Refsdal exploded approximately 9.3 billion years ago. The light from this explosion travelled to Earth along multiple paths, bent by the gravitational influence of dark matter. As each path took a different length of time to reach us, astronomers were able to observe the explosion on three separate occasions: in 1998, 2014, and 2015.

★ ★

Betelgeuse, a red supergiant star located 650 light-years away, is typically the tenth-brightest star in our night sky. It's so immense that if you were to pour 1,000 Earths into it every second, it would take over 100,000 years to fill. Betelgeuse is predicted to go supernova at any time within the next 100,000 years. When this happens, its core will collapse into a neutron star or black hole, and its outer layers will explode at 40,000 kilometres per second. Given Betelgeuse's close proximity to Earth, when it goes supernova, it will appear in our sky as bright as the full Moon and remain visible in broad daylight for three months.

Neutron Stars

A neutron star is an incredibly dense remnant left behind after a massive star explodes in a supernova. The gravitational forces driving this collapse are so intense that protons and electrons in the star's core are crushed together to form neutrons. During this event, atoms measuring one-tenth of a nanometre wide are compressed into a dense blob of neutrons just a few femtometres thick. That's equivalent to Earth shrinking down to the size of a city block.

★ ★

Neutron stars have a diameter of about 20 kilometres, yet they contain a mass roughly twice that of the Sun. This makes them the densest form of matter in the Universe.

★ ★

One tablespoon of a neutron star weighs as much as Mount Everest.

★ ★

If all the normal matter in the entire observable Universe were compressed to the density of a neutron star and placed where the Sun is, it would form a sphere extending to the orbit of Mars.

★ ★

A neutron star's equator is a perfect circle to within the width of a human hair.

★ ★

The tallest mountains on a neutron star are only a fraction of a millimetre high.

★ ★

A neutron star has such a strong gravitational field that you can see its entire surface from a single vantage point.

★ ★

A neutron star's gravitational pull is so powerful that if you were to drop an object from just one metre above its surface, it would hit the ground at a speed of 1,400 kilometres per second.

★ ★

If you dropped a marshmallow onto a neutron star, it would strike the surface with the energy of 1,000 hydrogen bombs.

★ ★

A neutron star's crust is believed to be made from a unique form of matter known as nuclear pasta. If confirmed to exist, nuclear pasta would be the strongest material in the Universe—ten billion times stronger than steel. Not only would nuclear pasta be incredibly strong, it would also be extremely heavy. In fact, the 100-metre-thick layer of nuclear pasta in a neutron star's crust would weigh more than 3,000 Earths.

★ ★

The core of a neutron star is subjected to such extreme pressure that it may contain an unusual substance called strange matter. This exotic material has a fascinating, yet terrifying, property: anything it touches is transformed into more strange matter. This means that if even a tiny fragment of strange matter were to collide with Earth, our entire planet would turn, atom by atom, into strange matter. Because of this, strange matter has been labelled 'the most dangerous substance in the Universe'.

★ ★

It's estimated that there are up to one billion neutron stars in the Milky Way alone.

★ ★

A rapidly rotating neutron star that emits radiation at regular intervals is known as a pulsar. The pulsar PSR J1748-2446ad is the fastest spinning object in the Universe. It rotates 716 times every second—so fast that its equator moves at one-quarter the speed of light.

★ ★

A neutron star with an exceptionally powerful magnetic field—about 1,000 times stronger than a typical neutron star—is known as a magnetar. If a magnetar were positioned halfway between Earth and the Moon, it would be powerful enough to erase the data from every credit card on the planet. If it were even closer—just 1,000 kilometres away—its magnetic field would strip the electrons from the atoms in your body, transforming you into a cloud of monatomic ions. This would effectively cause your body to dissolve.

★ ★

If the surface of a magnetar slips by as little as one millimetre, it can trigger a starquake that unleashes a powerful burst of radiation. Around 42,000 years ago, the magnetar SGR 1806-20, located on the far side of the Milky Way, experienced a starquake. In just one-tenth of a second, it released more energy than the Sun emits in 150,000 years. In 2004, the gamma rays from this explosion—the largest observed in our galaxy since a supernova witnessed in 1604—finally reached Earth. They compressed our planet's magnetic field, disabled satellites, and disrupted radio communication. Had this starquake occurred within ten light-years of Earth, it could have caused a mass extinction.

Gamma-Ray Bursts

Gamma-ray bursts are brief eruptions of high-energy light triggered by some of the most violent events in the Universe, such as neutron star collisions and the birth of black holes. These bursts are the most powerful explosions in the Universe, capable of releasing more energy in just a few seconds than the Sun will emit over its entire ten-billion-year lifetime.

★ ★

Gamma-ray bursts are extremely rare events, typically occurring only a few times in a galaxy every million years.

★ ★

Around 445 million years ago, Earth was plunged into a sudden ice age, resulting in the extinction of 85% of all species. Some scientists speculate that this catastrophic event was caused by a gamma-ray burst that took place less than 6,000 light-years away.

★ ★

On 19 March 2008, the gamma-ray burst 080319B was visible from Earth with the naked eye for 30 seconds. This burst originated 7.5 billion light-years away, making it the most distant recorded object ever observed without the aid of a telescope. The afterglow from this burst was the most intrinsically bright object ever witnessed by humans, shining 2.5 million times brighter than the brightest supernova.

Cosmic Rays

Cosmic rays are high-energy particles that travel through space near the speed of light. They come from various sources, including the Sun, supernovas, and even black holes. Despite being subatomic in size, a single cosmic ray can carry as much kinetic energy as a baseball thrown at 90 kilometres per hour.

★ ★

Cosmic rays are a type of ionising radiation that damages human tissue, degrades DNA, and contributes to a small percentage of cancer cases worldwide each year. The effects of cosmic rays on people living near sea level are typically minimal thanks to Earth's protective atmosphere. However, frequent airline travellers face increased exposure levels when flying at the altitude of commercial flights. Outside Earth's protective atmosphere, cosmic rays become an even greater threat. In fact, they are currently one of the major obstacles preventing humans from undertaking interplanetary travel.

★ ★

Astronauts aboard the International Space Station frequently see dots, streaks, and clouds, even when their eyes are closed. These visual phenomena are caused by cosmic rays interacting with their retinas.

★ ★

Cosmic rays can cause computers and smartphones to freeze or malfunction. The more powerful the computer, the greater the risk. A basic mobile phone with 500 kilobytes of memory might experience a cosmic ray-induced error once every 28 years. In contrast, a router farm with 25 gigabytes of memory, such as those used by internet providers, might experience one error every 17 hours.

★ ★

During the 2003 Belgian election, which utilised electronic voting, one candidate unexpectedly received an extra 4,096 votes. Following an investigation, it was concluded that a cosmic ray was most likely responsible for this anomaly. It's believed that the cosmic ray collided with the silicon in the election computer's microchip, creating a bit at position 13 in the computer's memory, leading to the extra votes.

★ ★

In 2008, a computer on board a Qantas commercial flight experienced a software error, causing the plane to plunge twice in rapid succession. This incident resulted in several serious injuries to the passengers and crew. The most likely cause of this computer malfunction was a cosmic ray.

Neutrinos

Neutrinos are tiny, nearly massless subatomic particles. They are produced in large quantities by various nuclear reactions, such as those occurring in stars and supernovas. They also arise terrestrially: in particle accelerators, nuclear reactors, and even through the radioactive decay of elements in everyday objects like bananas.

★ ★

The neutrino is the lightest known elementary particle. The next lightest known particle, the electron, is at least one million times as heavy.

★ ★

The neutrino is extremely small. In fact, if an atom were scaled up to the size of the Solar System, a neutrino would be the size of a golf ball.

★ ★

Neutrinos are often called 'ghost particles' because they rarely interact with matter. In fact, a neutrino could pass through a one-light-year-thick block of lead without interacting with a single atom.

★ ★

Most of the neutrinos detected on Earth are produced by nuclear reactions in the Sun. Every second, about 70 billion of these neutrinos pass through an area the size of your thumbnail. This means roughly 2.5 x 10^{21} neutrinos will sweep through your body over your lifetime. However, the likelihood of any single neutrino interacting with an atom in your body is extremely low—equivalent to flipping a coin 75 times and getting tails every time. When you balance the vast number of neutrinos passing through your body with the rarity of interactions, it translates to a 25% chance that a neutrino will interact with an atom in your body during your lifetime.

Telescopes

The Hubble Space Telescope not only transformed our view of the cosmos but also improved the early detection of breast cancer. When Hubble was launched in 1990, its images were blurry, so NASA developed advanced image processing techniques to sharpen them. This technology was later used to enhance the clarity of mammograms, enabling doctors to detect smaller calcifications, and thereby diagnose breast cancer earlier.

★ ★

In 2016, radio astronomers achieved the highest resolution in astronomical history by observing two regions of intense radiation just 20 kilometres apart, located near a star 6,500 light-years away. To put this level of resolution into perspective, it's equivalent to using a telescope on Earth to detect a flea on the surface of Pluto.

★ ★

Radio astronomers achieve exceptionally high resolution through a technique called very long baseline interferometry. It works by combining signals from radio telescopes that are located significant distances apart. This effectively creates a virtual telescope as large as the distance between the radio telescopes. Using this method, radio astronomers can simulate a telescope the size of Earth. If space-based radio telescopes are included in the network, they can simulate a telescope even larger than Earth.

★ ★

NASA is planning to build a one-kilometre-diameter radio telescope inside a crater on the far side of the Moon. If completed, it will become the largest filled-aperture radio telescope ever constructed.

★ ★

A liquid telescope uses a reflective liquid, typically mercury, as its primary mirror instead of solid glass. The liquid is spun in a container at a steady speed, so that it forms a smooth, parabolic surface ideal for focusing light. Liquid telescopes are about 95% cheaper to build than conventional telescopes; however, since their liquid mirror cannot be tilted, they are limited to looking straight up into the sky. The largest liquid telescope ever constructed was Canada's Large Zenith Telescope, which had a six-metre diameter.

★ ★

The James Webb Space Telescope is so sensitive to heat that it could theoretically detect the heat signature of a bee from 500,000 kilometres away. That's greater than the distance between Earth and the Moon.

★ ★

The mirrors on the James Webb Space Telescope are so smooth that if you enlarged them to the size of Australia, the largest bump would be just five centimetres tall.

★ ★

The James Webb Space Telescope doesn't orbit the Earth. It actually orbits the Sun at the second Lagrange point, 1.5 million kilometres away from the Earth.

★ ★

If an alien species living 80 million light-years away were to observe our planet through an ultra-powerful telescope, they would theoretically see dinosaurs roaming the Earth at this very moment.

Miscellaneous Astronomy

A civilisation only slightly more advanced than ours could use its star as a gravitational lens to detect the lights of New York City from as far as 500 light-years away.

★ ★

It's easier for a spacecraft to leave the Solar System than to reach the Sun. This is because everything on Earth is hurtling through space at 30 kilometres per second. To reach the Sun, a spacecraft must counteract this momentum and slow down by 30 kilometres per second, effectively cancelling out Earth's speed. In contrast, escaping the Solar System requires a much lower velocity—just 11 kilometres per second—making it a significantly easier task.

★ ★

The Oort cloud is a spherical shell of icy particles that defines the outermost boundary of our Solar System. Its distant edge stretches so far from the Sun that it may reach halfway to the Alpha Centauri star system.

★ ★

When you shine a torch at the sky, you release photons that will one day leave our galaxy.

★ ★

The Local Group, which includes the Milky Way and roughly 80 other galaxies, spans ten million light-years in diameter. Despite its vast size, it makes up only 0.00000000001% of the observable Universe.

★ ★

Our galaxy, along with 100,000 others, is being pulled towards a mysterious region of space called the Great Attractor. We are racing towards this gravitational anomaly at 600 kilometres per second; however, due to the expansion of the Universe, we will never actually reach it. The true nature of the Great Attractor remains a mystery because the Milky Way's galactic plane blocks our view of it, making it difficult to study.

★ ★

Space is not a perfect vacuum—there are an average of 0.3 atoms in every cubic metre of intergalactic space.

★ ★

The Universe contains voids that are so empty that if you teleported into one, you would just see darkness in all directions.

★ ★

The Bootes Void is a region of space that spans 330 million light-years and makes up 0.27% of the diameter of the observable Universe. It's often referred to as a supervoid because it contains just 60 galaxies, instead of the 2,000 galaxies that should exist in an area so vast. If the Milky Way had been located at the centre of the Bootes Void, humanity wouldn't have known about the existence of other galaxies until the 1960s.

★ ★

Einstein's Cross is a quasar—a bright galactic core powered by a supermassive black hole—located eight billion light-years from Earth. When viewed through a telescope, it appears as four distinct images arranged in the shape of a cross. This striking effect occurs because a galaxy, positioned directly between Earth and the quasar, bends the quasar's light through a process called gravitational lensing. As a result, the light from the quasar takes four different paths to Earth, creating the illusion of four separate quasars.

★ ★

The Tarantula Nebula is an extremely luminous nebula situated 160,000 light-years away. If it were as close to us as the Orion Nebula—which lies 1,500 light-years away—it would fill half our night sky and shine so brightly that it would cast visible shadows on Earth's surface.

★ ★

The Hercules-Corona Borealis Great Wall is a massive galaxy filament made up of a string of galaxy groups and clusters. It is the largest known structure in the Universe, spanning ten billion light-years across. To put this immense size into perspective, it's one-fifth the radius of the entire observable Universe.

BIOLOGY

Plants

Bamboo is the fastest-growing plant in the world. Some species grow at a rate of one millimetre every 90 seconds, which equates to 96 centimetres per day. This growth rate is so rapid that you can effectively see it growing in real time.

★ ★

Antarctic mosses are the slowest-growing plants on Earth, with some growing at a rate of just one millimetre per year. Due to this incredibly slow growth, they can be hundreds of years old, yet only a few centimetres tall.

★ ★

The seagrass Posidonia australis is the world's largest living organism by area. A single genetic specimen of this seagrass, found in Shark Bay, Western Australia, covers an area of 200 square kilometres, making it larger than the country of Liechtenstein.

★ ★

The scent of freshly cut grass is actually a chemical distress signal released by the grass.

★ ★

Carnivorous plants occasionally eat small animals, such as rats, frogs, and birds. If they're fed human flesh, they will eat that too.

★ ★

The Venus flytrap can count to five. When its trigger hairs are touched twice in quick succession, it snaps shut. If its hairs are touched three more times after it closes, it releases a digestive liquid to dissolve its prey. This unique counting mechanism ensures that it only snaps shut on live prey and not on inedible objects like raindrops and twigs.

★ ★

The Arabidopsis thaliana plant can perform division. To determine the optimal rate at which it should break down starch, it divides the total amount of starch it has accumulated by the corresponding time period. This calculation allows it to effectively manage its starch reserves.

★ ★

The tropical fern Ophioglossum reticulatum has 1,440 chromosomes—the most of any known plant. In comparison, humans have just 46 chromosomes.

★ ★

The Welwitschia mirabilis plant is known as a 'living fossil' because it has remained virtually unchanged for 200 million years, dating back to an era when dinosaurs first roamed the planet. It's composed of just two leaves, each measuring up to four metres long. These leaves can live for thousands of years; however, over time, they gradually become frayed by the relentless desert winds. The Welwitschia mirabilis is endemic to the Namib Desert, one of the driest places on Earth. To obtain the moisture needed for survival, it relies on a thick sea fog that gets blown up to 150 kilometres inland every few days.

★ ★

In 2022, researchers successfully grew plants in lunar soil. The soil, which had been collected by the Apollo astronauts, was glassy, metallic, and rich in minerals not commonly found on Earth. Despite these factors, the plants managed to grow. This result suggests that humans might be able to live on the Moon for extended periods by growing their own food in the lunar soil.

★ ★

Seeds determine the ideal time to sprout by balancing two factors: sprouting too early and being damaged by the cold weather, or sprouting too late and losing ground to competing plants. To assess this trade-off, they release two hormones: gibberellin, which encourages growth, and abscisic acid, which delays it. The seed's control centre, functioning much like a human brain, processes the returning hormonal signals to determine the perfect time to sprout.

★ ★

In nature, the line between plants and animals isn't always clear-cut. For instance, corals and sea anemones spend their entire life fixed in one place, and lack a heart, brain, and eyes—yet they are classified as animals. In contrast, some plants, like Venus flytraps and pitcher plants, display animal-like predatory behaviour by capturing and digesting insects for nutrition—yet they are classified as plants.

Plant Perception

In the early 2010s, plant ecologist Ernesto Gianoli was exploring the rainforests of southern Chile when he noticed two nearly identical plants growing side by side. One was the arrayan, a common plant in the region, while the other had the same leaves as the arrayan but a much thinner stem. Intrigued, Gianoli investigated further and discovered that the second plant was the Boquila trifoliolata. This plant has an extraordinary ability: it can change the shape, size, orientation, and vein pattern of its leaves to perfectly mimic other plants. One study found that it might even be able to mimic the leaves of plastic plants. What makes the Boquila trifoliolata even more fascinating is that it doesn't require physical contact to mimic a plant—it only needs to be nearby. The exact mechanism behind its mimicry remains a mystery, but one theory proposes that its lens-like cells act as rudimentary eyes, allowing it to see nearby plants and replicate their appearance.

★ ★

In 2020, researchers placed pea plants near a wooden support stick and observed how they adjusted their growth pattern over time. Remarkably, the plants were able to determine the stick's dimensions before they made physical contact with the stick. It's not known how they achieved this, although three hypotheses have been proposed. First, the plants may have employed a form of echolocation, emitting sonic clicks and interpreting the returning echoes to map their surroundings. Second, the plants' cells may have acted as primitive, light-sensitive structures, akin to rudimentary eyes, enabling them to visually perceive the stick.

Third, the plants may have detected the chemical cues released by the stick, essentially smelling its presence.

★ ★

In a 2016 study, researchers explored the risk-taking behaviour of pea plants by observing how they allocated resources. The plants had their roots split into two separate pots, so they were forced to choose which pot to prioritise. One pot contained a variable nutrient supply, while the other contained a constant nutrient level. When the conditions were unfavourable, the plants took a risk and opted to grow in the pot with the variable nutrient supply. In contrast, in favourable conditions, the plants became risk-averse and selected the pot with the constant nutrient level. This study demonstrates that plants can make informed decisions by evaluating risk factors, much like how humans do.

★ ★

A 2017 study found that when a plant 'hears' the sound of running water, it will grow its roots towards the water source. However, when a plant is exposed to the recorded sound of running water, it won't grow its roots towards the sound. This result suggests that plants can distinguish between real environmental cues and artificial ones.

★ ★

The silver thistle plant can predict rain hours in advance. It operates like a living barometer, so in dry conditions, it opens its petals wide, but when rain is approaching, it closes them tightly. This predictive ability persists even after the plant has died. Before the advent of modern weather forecasting, people would hang dead silver thistle on their doors to predict incoming rain.

★ ★

The Mimosa pudica is also known as the 'sensitive plant' because, when it's touched, it reacts instantly by folding its leaves inwards and drooping. It has developed this rapid movement as a defensive mechanism against herbivorous animals and insects.

★ ★

When the Mimosa pudica is given anaesthetic, it no longer folds its leaves in response to touch. Because anaesthetics impair consciousness in humans, this surprising reaction by the Mimosa pudica has led some scientists to propose the controversial idea that plants may possess a form of consciousness.

★ ★

When insects feed on cucumber, corn, tomato, or cabbage plants, the plants release specific chemicals into the air to attract the natural predators of the insects. The predators then land on the plants and consume the insects, preventing the plants from experiencing any further damage.

★ ★

Plants communicate with each other using clicks that resemble the sound of bubble wrap being popped. These clicks are as loud as human speech but occur at high frequencies. This means they can be heard by nearby plants, mammals, and insects, but are inaudible to human ears. Healthy plants typically emit fewer than one click per hour, while distressed plants can produce dozens of clicks over the same timeframe. Researchers have developed a machine learning program that can determine, with 70% accuracy, whether a plant's clicks indicate thirst or the threat of being cut.

Flowers

Wolffia globosa is the world's smallest flowering plant. It measures just 0.1 to 0.2 millimetres in diameter—about the same size as a grain of salt. Remarkably, the flowers it produces are microscopic, typically needing 20x magnification to be seen.

★ ★

Rafflesia arnoldii is the largest flower on Earth, measuring up to one metre in diameter and weighing as much as 11 kilograms. To attract pollinating flies, it emits a pungent odour, which closely resembles rotting flesh.

★ ★

The petals of the Diphylleia grayi flower turn transparent when they're exposed to rain, then return to their original white colour once they dry.

★ ★

Hydrangea macrophylla flowers are natural pH indicators. They turn blue if the soil is acidic and pink if the soil is alkaline.

★ ★

Placing a Viagra pill into a vase of freshly cut flowers will keep the flowers upright and unwilted for a week longer than normal.

★ ★

Violets steal your sense of smell. The unique scent of violets comes from a chemical called ionone, which briefly stimulates your scent receptors before temporarily disabling them. As a result, you can only smell violets for a few fleeting moments before their scent seems to vanish. To make their scent return, you simply need to take a few breaths.

Trees

There is no universally accepted definition of what constitutes a tree. Most definitions focus on specific characteristics, such as having a minimum height, woody stem, single trunk, or secondary growth. However, each of these proposed definitions excludes plants that most people would consider trees.

★ ★

Approximately 95% of a tree's mass comes from the air. In the 17th century, Belgian scientist Jan Baptist van Helmont demonstrated this with a famous experiment. He grew a tree in a pot of soil and found that after five years, the tree had gained a significant amount of mass, yet the soil's mass had hardly changed. Van Helmont correctly concluded that a tree's mass comes from something other than the soil. It turns out that a tree absorbs carbon dioxide from the air through photosynthesis and converts it into glucose. This glucose is then used to make cellulose, lignin, and other organic compounds that form the tree's structure. Through this process, a tree gains 95% of its mass from the air.

★ ★

Only 1% of a tree is actually alive. The remaining 99% is made up of non-living cells.

★ ★

Bonsai fruit trees produce regular-sized fruit.

★ ★

Redwoods are the tallest trees in the world, growing up to 116 metres in height. Despite their immense size, their root system is surprisingly shallow, extending just two to three metres deep into the ground. The reason they are so stable is because their roots stretch 30 or more metres horizontally.

★ ★

The dwarf willow is the smallest tree in the world, typically reaching only one to six centimetres in height.

★ ★

The largest tree in the world is a quaking aspen in Utah called Pando. While Pando resembles a forest of 47,000 individual tree stems, these stems are actually interconnected by a root system spanning 43 hectares—roughly the size of Vatican City. This means Pando is a single organism.

★ ★

Despite growing up to 60 metres tall, Cook pines usually lean towards the equator. This means that in the Northern Hemisphere they lean southwards, while in the Southern Hemisphere they lean northwards. The farther a Cook pine is from the equator, the greater the angle of tilt. On average, Cook pines lean at an angle of 8.55 degrees—twice that of the Leaning Tower of Pisa. However, one notable Cook pine in Adelaide, South Australia, leans northwards at an impressive 40-degree angle.

★ ★

There is evidence that trees sleep at night. A 2016 study of birch trees found that they enter a nightly sleep-like state, during which their branches relax and droop by as much as ten centimetres.

★ ★

Trees can help solve murders. When a human body decomposes after burial, it releases various nutrients into the surrounding soil. A tree growing near a buried body will absorb these nutrients, which can alter the colour and reflectance of its leaves. These subtle changes in a tree's leaves can be detected by specialised devices that analyse fluorescent signatures. With minor advancements in technology, trees may soon become a standard tool in murder investigations.

★ ★

Hura crepitans is often called the 'dynamite tree' due to its explosive seed dispersal mechanism. When its fruit pods ripen, they burst open, launching its seeds with such incredible force that they reach speeds of up to 250 kilometres per hour. This process propels the seeds as far as 100 metres, allowing the species to effectively spread and colonise new areas.

★ ★

Pine pollen contains all four major male sex hormones: testosterone, DHEA, androsterone, and androstenedione.

★ ★

In 2008, Japanese scientists sent cherry blossom seeds to the International Space Station, where they spent eight months in space. When the seeds were planted back on Earth, one grew into a tree that bloomed six years ahead of schedule and produced flowers with five petals, as opposed to its parent tree's 30 petals.

★ ★

There are more trees on Earth than stars in the Milky Way Galaxy. Earth is home to an estimated three trillion trees, while the Milky Way contains approximately 400 billion stars.

Fungi

Mushrooms communicate with each other using a 50 'word' vocabulary. These messages are transmitted by electrical impulses that travel along a network of underground filaments.

★ ★

Mushrooms are more closely related to humans than to plants. In fact, we share approximately 60% of our gene repertoire with mushrooms.

★ ★

Mushrooms and humans get infected by many of the same types of pathogens.

★ ★

Some mushroom species have more than 17,000 distinct sexes.

★ ★

Mushrooms are not green, so they don't photosynthesise. Instead, they obtain nutrients by breaking down organic material in their environment, such as dead leaves, wood, and other decaying matter. This is why they are commonly found in forests and wooded areas rich in this decaying material.

★ ★

The Armillaria ostoyae mushroom in eastern Oregon is the world's second-largest organism. This single specimen covers 9.6 square kilometres, making it larger than the territory of Gibraltar.

★ ★

Chorioactis geaster is an exceptionally rare mushroom found in only two places on Earth: small pockets in Japan and Texas. DNA analysis has shown that these two populations have remained genetically isolated for over 19 million years, evolving independently since their last common ancestor. It's not known why Chorioactis geaster has such an unusual distribution.

★ ★

The Ophiocordyceps unilateralis is commonly known as the zombie-ant fungus because it hijacks an ant's central nervous system to take control of its behaviour. Once infected, the ant becomes compelled to climb up a branch, bite into it, and remain attached until it dies. Following the ant's death, the fungus releases spores that spread to other ants, perpetuating the cycle of infection.

Agriculture

Three-quarters of the global food supply comes from just twelve species of plants and five species of animals.

★ ★

Just three foods—wheat, rice, and corn—provide more than half of the plant-derived calories consumed by people worldwide.

★ ★

Roughly 300,000 edible plant species exist on Earth, yet humans consume only 200 of them.

★ ★

At the start of the 19th century, approximately 80% of the US workforce was employed in the agricultural industry. Today, that figure has dropped to less than 2%.

★ ★

The vitamin and mineral content in fruits and vegetables has decreased significantly over the past century. For example, a 2018 study found that the calcium, magnesium, and iron levels in cabbage, lettuce, spinach, and tomatoes dropped by between 80% and 90% from 1914 to 2018. This sharp decline is primarily due to modern agricultural practices, which have depleted essential nutrients from the soil.

★ ★

Brussels sprouts, cabbage, kale, broccoli, kohlrabi, and cauliflower all come from the same plant: Brassica oleracea.

★ ★

Avocados shouldn't exist. Most plants survive because animals eat their seeds and spread them through excretion. However, avocados have a unique problem: their seeds are so large that only the great prehistoric mammals, like giant sloths and gomphotheres, could swallow and disperse them. These animals went extinct around 13,000 years ago. Scientists don't know how avocados survived the thousands of years between the extinction of these megafauna and the start of human cultivation.

★ ★

Cauliflower grows so fast—up to three centimetres per day—that you can actually hear it growing. As its popcorn-like heads expand, they rub against each other, producing a faint yet audible sound known as the 'cauliflower creak'.

★ ★

An ear of corn always has an even number of kernel rows—except in rare cases where significant stress has disrupted its development.

★ ★

The apples you bought from the supermarket today were picked off a tree approximately one year ago.

★ ★

Most citrus fruits don't exist in the wild. There are just four original citrus species: the mandarin, pomelo, citron, and papeda. Every other citrus fruit we know today, including the orange, lemon, and lime, is a hybrid that was created by cross-pollinating these four foundational fruits.

★ ★

Pink bananas exist—and when they ripen, they naturally peel themselves.

★ ★

Have you ever wondered why bananas are curved? When bananas are young, they grow straight downwards because gravity pulls them towards the ground. However, as they mature, they grow upwards, turning towards the sunlight. This upwards growth against gravity, called negative geotropism, gives them their distinctive curved shape.

★ ★

Bananas naturally produce antimatter. Due to the decay of potassium-40, they emit an average of one positron every 75 minutes.

Viruses

Every day, viruses kill half of the ocean's bacterial population and 20% of the ocean's total living biomass.

★ ★

A single millilitre of coastal water from the ocean's surface—enough to fill an eyedropper—contains up to ten million viruses.

★ ★

The ocean is home to one nonillion viruses—that's one followed by thirty zeros. If these viruses were laid end to end, they would span ten million light-years.

★ ★

Viruses are so tiny that they are smaller than the wavelength of visible light. This means they have no colour and cannot be directly observed using a light microscope.

★ ★

If you gathered all the HIV particles in existence—from the 40 million people currently infected worldwide—they would only fill a single teaspoon.

★ ★

Approximately 500 million rhinoviruses—the virus responsible for the common cold—could fit on the head of a pin.

★ ★

Over just four months, the 1918 influenza pandemic killed more people than World War I.

★ ★

During the late 1980s, the Soviet Union maintained a biological warfare stockpile that included over 18 tonnes of smallpox. At the time, the combined weight of all the other human viruses in the world was less than nine tonnes. This meant that the Soviet Union's biological warfare stockpile contained the majority of all human viruses on Earth, measured by weight, volume, and number.

Bacteria

If all the bacteria on Earth were laid end to end, they would stretch for ten billion light-years—roughly one-tenth the diameter of the observable Universe. However, this bacterial thread would be difficult to see, as it would be 75 times thinner than a human hair.

★ ★

The oceanic bacterium Pelagibacter ubique is the most abundant organism on Earth. With an estimated 2×10^{28} of these bacteria in existence, they outnumber the stars in the observable Universe by a factor of 10,000. Pelagibacter ubique is so common that its combined weight exceeds that of all the fish in the ocean. Despite its prevalence, the bacterium was only discovered in 2002 and is yet to be officially named.

★ ★

All the bacteria on Earth collectively weigh over 1,000 times more than all the humans.

★ ★

If all the bacteria in the world suddenly vanished, humans wouldn't notice for about a week. However, within a year, society would collapse.

★ ★

Under ideal conditions, the stomach bacterium E. coli doubles its population every 20 minutes. Starting with a single cell, after 20 minutes, it becomes two; after 40 minutes, it becomes four; and after one hour, it becomes eight. Given unlimited resources, after just 36 hours, the number of E. coli cells would be so vast that they would cover the entire surface of the Earth to a depth of 30 centimetres. And after only 89 hours, the E. coli would have produced more offspring than there are protons in the observable Universe.

★ ★

Thiomargarita magnifica is the world's largest known bacterium. It typically grows to around one centimetre in length, making it easily visible to the naked eye. To put its size into perspective, it's 5,000 times larger than a standard bacterium—the equivalent of a human encountering another human as tall as Mount Everest. Thiomargarita magnifica is extremely rare; it has only ever been discovered in the sediment-rich waters of a single mangrove swamp in Guadeloupe.

★ ★

Magnetotactic bacteria have the unique ability to sense Earth's magnetic field and use it for navigation. Leveraging this trait, scientists have utilised these bacteria for several practical applications, such as targeted drug delivery within the human body and the removal of toxic metals from water.

★ ★

Pseudomonas syringae is a bacterium with a remarkable ability: it can freeze water that's at above-freezing temperatures. When these bacteria come into contact with water, they provide nucleation, rearranging the surrounding water molecules, turning them into ice. Pseudomonas syringae are so effective at making ice that they are routinely added to snow machines at ski resorts to enhance snow production.

★ ★

The wreck of the Titanic is being eaten away by a species of bacteria called Halomonas titanicae. These iron-consuming microorganisms are gradually breaking down the ship's structure, leading some scientists to predict that most of the Titanic will disappear entirely within the next few decades.

★ ★

Pound-for-pound, the gonorrhoea bacterium is the strongest creature on Earth. It can pull a load 100,000 times its own body weight—the equivalent of a person dragging the Eiffel Tower.

★ ★

Most of your body is not human. In fact, a typical adult is composed of 38 trillion bacterial cells and just 30 trillion human cells.

★ ★

At every moment, you are surrounded by a personalised cloud of bacteria, called a microbial aura, extending 90 centimetres from your body. This cloud is made up of millions of bacteria shed from your skin, hair, and breath. These bacteria are uniquely yours, determined by factors such as your diet, hygiene, and environment. While most of these bacteria are harmless or even beneficial, some have the potential to spread infection.

★ ★

The flagellar motor is a natural motor used by many bacteria to propel themselves through liquid environments. Despite being one millionth the size of a grain of sand, it can swim hundreds of its body lengths every second and change direction faster than a mosquito flaps its wings. The flagellar motor spins at up to 100,000 revolutions per minute and boasts an energy conversion efficiency of nearly 100%. This means that it rotates with a higher speed and greater efficiency than any human-made device.

★ ★

In 2024, scientists created genetically engineered bacteria capable of working together to solve problems. When presented with yes-

or-no questions through chemical signals, the bacteria respond by secreting proteins to indicate their answer: red for *"no"* and green for *"yes"*. These bacteria can recognise vowels, identify prime numbers, and even calculate the maximum number of pizza slices produced by a given number of straight cuts.

Microorganisms

All the microorganisms on Earth weigh more than all the animals combined.

★ ★

A single teaspoon of soil contains approximately one billion microorganisms.

★ ★

An estimated one trillion species of microorganisms inhabit Earth, yet we have only discovered 0.001% of them.

★ ★

Diatoms are the most important microorganisms that most people have never heard of. These tiny algae generate 20% to 40% of the oxygen we breathe, making them the largest primary producers of oxygen on the planet.

★ ★

The polar ice caps are melting and unleashing prehistoric microorganisms that have been trapped for millennia. It's estimated that four sextillion of these microorganisms are being released annually. Scientists are concerned that some may be pathogenic viruses and bacteria that could revive and threaten human health. There is a precedent for this concern. In 2016, rising temperatures caused part of the Siberian permafrost to melt, resulting in the release of anthrax spores that killed one person and infected dozens more.

★ ★

In 2010, scientists conducted an experiment in which they arranged nutrients in the same pattern as the stops on the Tokyo subway system. They then introduced slime mould into the experiment. Within just one day, the slime mould created a network that was nearly identical to Tokyo's real subway system. The Tokyo subway system was meticulously designed by engineers to be as efficient as possible. The slime mould designed the same network—without a brain.

★ ★

Scientists have discovered microorganisms over 100 million years old that are still alive today. These microbes, buried in the sediment beneath the seafloor, have achieved a state of near-immortality by surviving on virtually no energy.

★ ★

Some biologists speculate that a shadow biosphere might exist on Earth. This would consist of a group of undiscovered microbial lifeforms with an entirely different biochemistry from what we're used to. These lifeforms would be invisible to us because our methods for detecting microbial life are based solely on the biochemistry we know. If scientists were ever to discover a shadow biosphere, it would suggest that life began independently on Earth more than once, raising the possibility that life is abundant throughout the Universe.

Life

Your very existence is a miracle. The chance of your parents meeting was roughly one in 20,000. The probability of that meeting resulting in children was about one in 2,000. The odds of the specific sperm fertilising the specific egg that created you was a one in 400 quadrillion event. Tracing back your direct ancestors, the odds of your lineage remaining unbroken for over 150,000 generations of human existence was roughly one in $10^{45,000}$. The likelihood of the right sperm meeting the right egg across all of these generations was approximately one in $10^{2,640,000}$. Each step in this process had to occur for you to exist. When you combine these probabilities, the chance of you being born comes out to one in $10^{2,685,000}$. To put this figure into perspective, it's equivalent to two million people each rolling a trillion-sided die and all landing on the same number—say 459,638,936,251.

★ ★

All known life on Earth has a carbon-based structure. This is because carbon atoms can form bonds with up to four other atoms simultaneously, enabling the creation of complex molecular chains, like DNA and proteins, which are essential for life as we know it. However, some scientists have proposed that extraterrestrial life might have a completely different atomic structure—one that is based on silicon. This is because, like carbon, silicon is extremely abundant throughout the Universe and can bond with up to four other atoms simultaneously.

★ ★

In 1952, Stanley Miller and Harold Urey conducted a groundbreaking experiment to determine whether the organic compounds essential for life could have arisen spontaneously on Earth. They simulated the conditions of the early Earth by sealing methane, ammonia, hydrogen, and water in a glass apparatus. They then introduced sparks of electricity to replicate lightning. After letting the experiment run for just one week, they successfully created amino acids—the molecular building blocks of life. The results of this experiment provide strong evidence that life on Earth could have arisen from non-living matter.

★ ★

In 2020, NASA collected samples from an asteroid named Bennu and brought them back to Earth for analysis. These samples contained 14 of the 20 amino acids present in life, along with all five nucleobases—the fundamental molecules that form RNA and DNA. This finding provides evidence that asteroids may have delivered the building blocks of life to our planet during its early history.

★ ★

Scientists have recently discovered just the fourth known instance in Earth's history of primary endosymbiosis—a phenomenon where two organisms merge to form a single lifeform. The first occurrence of this phenomenon led to the emergence of all complex life. The second resulted in the proliferation of plants across the planet. The third gave rise to an organelle capable of photosynthesis. The latest discovery is of a marine alga that engulfed a bacterium 100 million years ago. Over time, these two organisms evolved into a single entity capable of converting atmospheric nitrogen into ammonia. It remains to be seen whether this latest instance of primary endosymbiosis will have any significant global implications.

Chirality

Amino acids and sugars have a unique property known as chirality, or 'handedness'. When these molecules are created in a laboratory, they naturally form a 50-50 mixture of left-handed and right-handed versions. However, on Earth, all amino acids that form life are left-handed, while all sugars that form life are right-handed. The opposite hands of these molecules don't constitute any known biological lifeforms. The reason why life is composed exclusively of left-handed amino acids and right-handed sugars remains one of the greatest mysteries in science.

★ ★

In theory, it should be possible to create a mirror version of life—one where amino acids are right-handed and sugars are left-handed. Within this ecosystem, mirror animals could only survive by consuming reflected food composed of mirror amino acids and sugars. Moreover, viruses would also need to be mirrored versions, since they can only infect hosts that have matching molecular structures.

★ ★

Imagine a scenario where scientists engineered a mirror cyanobacteria capable of photosynthesis. If even a single droplet of this modified bacteria were released into the ocean, it would consume every available nutrient, starving all other marine life. This would cause the ocean ecosystem to collapse, triggering a global famine that would plunge the planet into an ice age.

★ ★

If mirror bacteria were ever created, they would likely evade the immune systems of humans, animals, and plants. This is because immune systems have evolved to identify specific molecular structures, and the mirror bacteria would have a reversed structure that isn't recognised. As a result, the mirror bacteria would potentially cause widespread, severe infections in humans. This situation would be catastrophic, since the many existing drugs designed to target standard molecular structures would be ineffective against the mirror bacterial infections.

★ ★

Some scientists have begun efforts to create mirror bacteria; however, it's estimated that developing such organisms will take at least a decade. Meanwhile, other scientists have called for a complete ban on all research aimed at producing mirror life, due to the existential threat it poses.

Genealogy

Choose any organism—from the bacteria in your yogurt to the lettuce in your garden. If you travel far enough back in time, you'll discover that you share a common ancestor with that organism.

★ ★

Most of your ancestors weren't humans—in fact, most weren't even animals.

★ ★

Your 200 millionth great-grandmother was a fish. She provided you with some of the genes that make up your DNA. Your arms and legs are just more advanced versions of the two pairs of fins that she used to swim through the ocean.

★ ★

Y-chromosomal Adam is the most recent man from whom all humans are descended. He lived approximately 200,000 years ago and is considered your 11,000th great-grandfather.

★ ★

Counterintuitively, you have twice as many female ancestors as male ancestors. This is because, throughout history, 80% of women have reproduced, while just 40% of men have reproduced.

★ ★

If you don't have children, you are the first person in your direct lineage not to reproduce since the dawn of time.

★ ★

If a woman doesn't have a daughter, she ends a streak of her direct ancestors having daughters that dates back to the beginning of humanity.

★ ★

All humans alive today are 50th cousins or closer.

★ ★

Throughout history, 80% of all marriages have been between second cousins or closer.

★ ★

Full siblings share anywhere from 37.4% to 61.7% of their DNA. This variation occurs because siblings inherit different combinations of genes from their parents due to the random nature of genetic recombination.

★ ★

You don't inherit DNA from all of your ancestors. In fact, 28% of your seventh great-grandparents didn't contribute at all to your genetic makeup.

★ ★

Twins can have different fathers due to a rare phenomenon called heteropaternal superfecundation. This occurs when a woman releases two eggs during ovulation, and each egg is fertilised by sperm from a different man within a short timeframe. An example of this phenomenon occurred in 2008, when Texas woman Mia Washington gave birth to twins. Paternity tests revealed that each twin had a different father.

★ ★

In extremely rare cases involving genetic disorders, identical twins can be different genders.

Human Genetics

The human genome is composed of 3.2 billion letters, which if printed, would fill a stack of paperback books reaching 61 metres high. Remarkably, nearly all of the body's 30 trillion cells contain an entire copy of this human genome.

★ ★

It's extremely rare for a virus to integrate its genes into its host organism's DNA. However, over the vast span of evolutionary history, these rare incidents have accumulated. As a result, approximately 8% of the human genome is now made up of viruses that infected our primate ancestors millions of years ago. This ancient viral DNA plays a dual role in human health: it protects us from certain illnesses, while simultaneously increasing our susceptibility to cancer and other diseases.

★ ★

Only 1% of the human genome is responsible for coding proteins. The remaining 99%—once dismissed as 'junk DNA'—contains no coding sequences whatsoever. However, scientists have recently discovered that this so-called junk DNA is far from useless: it serves vital functions, such as controlling when and how genes are turned on or off.

★ ★

If all the DNA produced by a human over a lifetime were laid end to end, it would span more than one light-year.

★ ★

You can observe your own DNA by performing a simple experiment. Start by adding a tablespoon of salt to 500 millilitres of water, then stir until the salt dissolves. Next, take about 50 millilitres of the saltwater solution and gargle it for one minute to collect cells from your mouth. After gargling, spit the solution into a clear glass. Add two drops of liquid detergent to the glass and gently stir the mixture. Next, slowly pour 100 millilitres of cold rubbing alcohol into the glass. Wait for three minutes and you'll notice stringy clumps forming—this is your DNA!

★ ★

During the hours and days following a person's death, hundreds of their genes turn on. This includes some genes that had been dormant since the person was a fetus.

★ ★

All human traits are at least partially heritable. This even includes unusual traits, like smartphone screen time and typing font preference.

★ ★

Complex human traits are shaped by thousands of genetic variants. Each of these variants individually has close to no effect; however, their cumulative impact is substantial. For example, thousands of genetic variants affect a person's IQ, yet each of these increases or decreases IQ by just a tiny fraction of a point.

★ ★

Having six fingers on a hand is a dominant trait in humans.

★ ★

Sugar is healthy for some people. In Greenland, 2% to 3% of the population has a genetic mutation that prevents their body from absorbing glucose. Instead, their gut bacteria convert sugar into healthy compounds. For people with this mutation, drinking a soft drink provides nearly as many health benefits as eating broccoli.

★ ★

Chimera is a rare condition in which a person has two different sets of DNA. This phenomenon can occur naturally when a developing fetus absorbs its twin during pregnancy. It can also arise from medical procedures, such as organ transplants, blood transfusions, or stem cell transplants. Interestingly, most people with chimera are unaware of their dual genetic makeup. A notable example of this is the 2002 case of Lydia Fairchild. She was accused of kidnapping her own children after DNA tests showed that she had a different genetic profile from theirs. It was later discovered that she was a chimera.

★ ★

Fetal microchimerism is a phenomenon in which fetal cells remain in the mother's blood and tissue long after pregnancy. These leftover cells can linger in the mother's body for decades, sometimes even for the rest of her life. The condition is nearly universal, affecting almost every woman who has been pregnant. For example, a 2012 study found that 63% of women had traces of male DNA in their brain as a result of having given birth to a son.

★ ★

Microchimeric cells appear to have their own evolutionary goals, such as boosting their chance of survival and enhancing their ability to replicate. Moreover, their presence can impact the host's health, behaviour, and the success of future pregnancies.

★ ★

A 2015 study found that the children of Holocaust survivors—who had not directly experienced trauma themselves—exhibited the same DNA modifications related to regulating stress hormones as their mothers. This discovery suggests that traumatic events can leave a biological imprint that potentially affects the health of future generations.

Genetics—General

In a 2013 study, researchers conditioned mice to associate the scent of cherry blossom with fear. Remarkably, when the descendants of these mice were later exposed to the same scent, they exhibited a similar fear response, even though they had never encountered the scent themselves. This finding suggests that certain experiences can leave a biological imprint that influences future generations.

★ ★

In a 2018 study, researchers successfully transferred 'memories' between snails. They began by training a group of snails to develop a defensive response to electric shocks. Next, they extracted RNA from these trained snails and transferred it to a group of untrained snails. Remarkably, the untrained snails exhibited the same sensitised defensive reaction as the snails that had been exposed to the shocks. This result suggests that memory-like information can be transferred through RNA.

★ ★

The C-value paradox is the observation that an organism's complexity does not correlate with its genome size. For example, humans, despite being biologically complex, have a genome with 3.2 billion base pairs. In contrast, the single-celled amoeba Polychaos dubium has a genome over 200 times larger, with 670 billion base pairs.

★ ★

A tomato has more genes than a human.

★ ★

Genes are typically passed vertically from parent to child. However, due to a process known as horizontal gene transfer, they can also be transferred between organisms or even across different species. Examples of horizontal gene transfer include gonorrhoea bacteria acquiring human DNA, sea slugs inheriting genes from the algae they consume, and smelt fish adopting genes from herring. Some scientists even speculate that plants may have transferred some of their genes to humans through horizontal gene transfer.

★ ★

Approximately 500 million years ago, a significant evolutionary step occurred when genes were passed from soil bacteria to algae through horizontal gene transfer. This genetic transmission contributed to plants moving from water to land.

★ ★

In a 1995 experiment, researchers took the mouse gene PAX6, which controls eye development in mice, and inserted it into the DNA of a fruit fly larva. Surprisingly, the gene didn't produce mouse eyes on the fly. Instead, it activated the fly's own developmental pathways, creating fully functional fruit fly eyes. This result suggests that different species share the same fundamental genetic instructions for development.

★ ★

Squid, octopuses, and other cephalopods can edit their own RNA. This unique trait allows them to quickly adapt to the changing seasonal temperatures in the ocean.

★ ★

In 2019, scientists created the world's first fully synthetic living organism: an artificial version of the stomach bacterium E. coli. The synthetic E. coli's genome was made up of four million base pairs—a complexity so vast that if it were typed out would fill 970 A4 pages.

Brain Organoids

Over the past decade, scientists have made significant progress in using stem cells to grow miniature brains called brain organoids. These blobs of human brain tissue provide a simplified model to study brain development, test drugs, and research neurological disorders. While brain organoids can respond to certain stimuli, such as light, they're too simple to feel pain or experience awareness. However, as scientists rapidly enhance their complexity, there are ethical concerns over whether brain organoids may one day develop some form of consciousness.

★ ★

If you take brain organoids representing different regions of the human brain and leave them in a petri dish overnight, they will spontaneously form connections with each other. These connections aren't random—the same brain regions that are interconnected in the human brain will also naturally connect in brain organoids.

★ ★

In 2021, researchers growing brain organoids in a laboratory observed that the organoids spontaneously developed light-sensitive, eye-like structures called optic cups. These structures began to emerge 30 days into the growth process and reached maturity after 50 days. This mirrors the retinol development timeline in human embryos.

★ ★

The longer that a brain organoid is cultivated, the more advanced it becomes. In fact, brain organoids cultivated for six months or longer exhibit electrical activity patterns similar to those found in premature babies.

★ ★

Brain organoids currently lack a vascular system, so they're unable to efficiently absorb oxygen and nutrients. As a result, they usually die from hypoxia once they reach about eight millimetres in diameter—typically after one year of growth. However, recent advancements may soon enable scientists to equip brain organoids with a vascular system, allowing them to grow indefinitely. It's not known how advanced brain organoids will become once they grow larger.

★ ★

Brain organoids are trained with dopamine—a neurotransmitter that stimulates them to form new pathways and connections. This mirrors the way that real human brains learn.

★ ★

Organoid intelligence is an emerging field in which brain organoids are connected to a conventional computer network. To date, these hybrid systems have successfully performed various tasks, including recognising speech, playing computer games, and solving simple mathematical problems. Organoid intelligence computer systems offer several potential advantages over traditional computer systems—they may learn faster, process more complex information, and operate with greater energy efficiency.

Miscellaneous Biology

Evolution can happen overnight. Larger organisms experience gradual evolutionary changes over long periods. In contrast, microorganisms can evolve quickly because they reproduce by splitting into two as frequently as every few minutes. This means that if a mutation gives a microorganism a reproductive advantage, it can rapidly spread through the population, becoming dominant within a single day.

★ ★

Multicellular life has evolved independently on at least 25 distinct occasions.

★ ★

An egg yolk is technically a single cell.

★ ★

The Caulerpa taxifolia seaweed is the world's largest single-celled organism. Despite growing up to 30 centimetres in length, it's composed of just one enormous cell.

★ ★

The Fibonacci sequence is a number series where each term is the sum of the preceding two. It goes 0, 1, 1, 2, 3, 5, 8, 13, 21, 34, and so on. The sequence often appears in nature—for instance, if you count the number of spirals running in either direction on a pine cone, it will be an integer in the Fibonacci sequence 97% of the time.

★ ★

A 1992 analysis of 650 plant species found that over 90% contained Fibonacci spirals.

★ ★

Fibonacci spirals weren't always so prevalent. A 2023 study of ancient plants dating back hundreds of millions of years found that Fibonacci spirals were once a rarity. Over time, they became far more common, suggesting that the pattern may have evolved as plants developed more efficient ways to grow.

★ ★

American microbiologist Maurice Hilleman developed more than half of the routine vaccines used today.

★ ★

In 1962, an anonymous Swedish woman underwent an elective abortion. The cell lines from her fetus were used to create vaccines against chicken pox, shingles, rubella, adenovirus, measles, mumps, poliovirus, hepatitis A, and rabies. Over the years, this single cell line has produced billions of vaccine doses, saving more than ten million lives worldwide.

★ ★

In 1951, an American woman, Henrietta Lacks, passed away from cervical cancer. The cancer cells extracted from Lacks' cervix were unlike any human cells scientists had ever seen. They behaved like a microbial species, doubling in number every 20 to 24 hours. Since Lacks' death, more than 50 million tonnes of her cells have been grown, distributed, and utilised in over 75,000 scientific studies. These cells have been instrumental in developing vaccines for polio and COVID-19, as well as in advancing AIDS and cancer research.

CHEMISTRY

Chemicals

The inorganic compound nitrogen triiodide is so unstable that even the light touch of a mosquito landing on it will cause it to explode.

★ ★

Azidoazide azide has been labelled 'the most explosive chemical compound ever created'. It's so unstable that scientists can't measure its properties without triggering an explosion. Even minor disturbances, like touching it, moving it, and obtaining an infrared spectrum of it, cause it to detonate.

★ ★

The chemical compound diazomethane is so unstable that even the slightest scratch on the glassware used to store it can trigger an explosion.

★ ★

Per- and polyfluoroalkyl substances (PFAS) are a group of synthetic chemicals, developed in the 1930s, that have become a major health and environmental hazard due to their toxicity and persistence. Known for their resistance to oil, water, heat, and grease, they are found in various everyday items, like non-stick pans, waterproof clothing, and food packaging. They are highly toxic—even at parts per quadrillion—and exposure is linked to a range of serious health conditions, including cancer, endocrine disorders, and developmental delays. Alarmingly, virtually every person on Earth has traces of PFAS in the blood, and these levels are only rising over time. PFAS don't break down naturally, so they linger in the environment indefinitely, earning them the name 'forever chemicals'. Due to this persistence, they are accumulating in the air, water, and soil, posing long-term risks to the ecosystem.

★ ★

Bandages may be harmful to your health. A 2024 study found that 65% of bandages on the market contain detectable levels of PFAS chemicals. Concerningly, many of these bandages have PFAS on their absorbent pad—the part that directly touches open wounds. This means that the chemicals can be directly absorbed into your body.

★ ★

In general, there is no such thing as a toxic chemical; there are only toxic doses. Virtually all chemicals are safe at low doses, while all chemicals are toxic at high doses. There are just a few exceptions to this rule—certain heavy metals, such as lead and mercury, can be toxic in just trace amounts. Furthermore, some substances, including carcinogens, have no completely safe dose, as even a single molecule can theoretically cause damage.

★ ★

A chemical's properties are determined by the other chemicals it's bound to. For example, sodium is highly reactive and will explode if it comes into contact with water, while chlorine is extremely toxic at anything above a minuscule dose. However, when sodium and chlorine are combined, they form sodium chloride, commonly known as table salt. This compound has neither the explosive reactivity of sodium nor the toxicity of chlorine. When a substance contains a 'dangerous' chemical, it's important to check whether it's bound to another chemical, rendering it safe.

Elements

Astatine is the rarest naturally occurring element on Earth. There are only about 25 grams of it present in all of Earth's crust at any one time—equivalent to the weight of five US nickels. Astatine is so rare that no solid sample has ever been observed because any visible amount would instantly vaporise due to the heat generated by its intense radioactivity. As astatine is so difficult to produce, scientists still don't know its basic properties.

★ ★

Just one gram of polonium—an amount the size of a single grain of rock salt—is enough to kill 50 million people.

★ ★

Oganesson, the heaviest element on the periodic table, is incredibly difficult to produce. Scientists have only ever created five atoms of it, each of which lasted for less than one millisecond. Remarkably, these might be the only five atoms of oganesson to have ever existed in the entire Universe.

★ ★

Tellurium-128 has the longest known half-life of any radioactive isotope, at 2.2 x 10^{24} years—equivalent to 160 trillion times the current age of the Universe.

★ ★

Approximately 16% of the elements on the periodic table exist exclusively in laboratories on Earth and are not found naturally anywhere in the Universe.

★ ★

Roughly 10% of your body weight comes from hydrogen that formed just after the Big Bang.

★ ★

You are, quite literally, made of stardust. Most of the elements that make up your body were forged billions of years ago within the heart of a star. Many of these elements have journeyed through multiple supernova explosions before becoming part of who you are today.

★ ★

Much of the iodine in your body was formed during the collision of two neutron stars.

★ ★

Just six elements—hydrogen, oxygen, carbon, nitrogen, calcium, and phosphorus—make up more than 99% of the atoms in your body.

★ ★

An iPhone is composed of more elements than a human. An iPhone is made up of around 75 elements, representing about two-thirds of the periodic table, whereas a human is made up of just 30 elements.

★ ★

The americium used in ionisation-type smoke detectors is the only synthetic element found in the household. A new smoke detector contains around 0.29 micrograms of americium-241—equivalent to roughly 1/3,000th the weight of a grain of sand. This tiny amount of americium ionises the air in the detector's chamber, creating a steady electrical current. When smoke particles enter the chamber, they disrupt this current, triggering the alarm.

★ ★

The Swedish village of Ytterby is home to just 3,000 people, yet it's the world's single greatest source of elemental discoveries. Four elements—yttrium, ytterbium, terbium, and erbium—are named after the village, while another five elements—gadolinium, holmium, scandium, tantalum, and thulium—were first discovered there.

Helium

Helium is the only element discovered in space before it was discovered on Earth. During the total solar eclipse of 1868, astronomers Pierre Janssen and Norman Lockyer independently observed a distinct yellow spectral line emanating from the Sun, which was later identified as helium. This element was named 'helios', after the Greek word for 'Sun'. Despite its late discovery, helium is remarkably abundant, accounting for one in every ten atoms in the Universe.

★ ★

Helium is formed deep within the Earth over hundreds of millions of years due to the natural radioactive decay of certain elements. Over time, it slowly rises through the Earth's crust, eventually becoming trapped in natural gas reservoirs, which we extract. Although countless articles claim that the world is on the brink of running out of helium, this is not the case. While it's true that we are depleting our current helium reserves, there is a vast supply of helium within the Earth. If we wish to obtain more helium, we need to build new reserves.

★ ★

The helium inside a helium balloon is destined for space. Helium atoms are the smallest atoms on the periodic table, so they easily slip through the tiny gaps in a balloon's material. Once they leak out, they rise through Earth's atmosphere until they escape into space.

★ ★

When you inhale a lighter-than-air gas, like helium, the sound waves travel faster through your vocal cords, causing your voice to sound squeaky. In contrast, when you inhale a denser-than-air gas, like sulphur hexafluoride, the sound waves travel more slowly through your vocal cords, causing your voice to sound deeper.

★ ★

Hospitals use helium to keep the powerful magnets in their MRI machines extremely cold. In fact, one-third of the world's total helium supply is used for this purpose. Due to helium's scarcity and high demand, a typical mid-sized hospital spends roughly US$20,000 annually on liquid helium.

★ ★

When liquid helium is cooled to just a few degrees above absolute zero, it transforms into a superfluid with extraordinary properties. One such property is that it will seep through the molecule-thin cracks in the walls of the container it's being housed in. Another property is that it will creep up the sides and over the edges of an open container until it's empty. A final property is that it will flow without friction, so it will move indefinitely. This means that it can be stirred, then left for one million years, and it will still be moving. These unique properties of superfluid helium are the result of quantum mechanics. However, unlike most quantum effects, they are visible to the naked eye.

Metals

The metallic smell of coins doesn't come from the metal—it comes from you. Freshly minted coins don't have a noticeable scent. However, as they circulate from person to person, they pick up oils from human skin. These oils react with the iron ions in the coins to create a chemical called 1-octen-3-one, which is responsible for the familiar 'coin smell'. This chemical has an extremely low odour threshold—around 0.01 parts per billion—meaning humans can detect it in minute concentrations.

★ ★

Gallium has such a low melting point—just 29.76 degrees Celsius—that it will melt from a solid metal to a liquid on the palm of your hand.

★ ★

When magnesium burns, the ash it produces weighs more than the original magnesium. This occurs because the magnesium reacts with the oxygen in the air to form magnesium oxide—a compound with a greater mass than pure magnesium.

★ ★

When two pieces of the same type of metal come into contact in the vacuum of space, they fuse, forming a single, unified piece of metal. This phenomenon is known as cold welding.

★ ★

Nitinol is an alloy composed of nickel and titanium that's renowned for its ability to return to its original shape after it's bent or deformed. This shape-memory property makes it incredibly useful in a wide range of applications. For example, it's used to make self-healing glasses frames that regain their shape after being bent. It's also used in orthodontic braces to gently straighten teeth over time. NASA has even harnessed its properties to build airless, shape-memory tyres for its lunar and Martian rovers.

Gold

Gold and copper are the only two metals on the periodic table that are not silver in colour.

★ ★

Gold's distinctive yellow colour is due to the effects of Einstein's theory of special relativity. Gold has large atoms, so its nucleus has a strong pull on its electrons, causing them to move at over half the speed of light. At such high speeds, relativistic effects—like increased mass and length contraction—come into play. These effects change the way gold absorbs light. So, instead of absorbing ultraviolet light, like most metals, it absorbs blue light. As a result, its complementary colour is reflected—the striking yellow we are familiar with.

★ ★

Gold nanoparticles are not a vibrant yellow colour. Instead, they come in a variety of other hues, including red, orange, and blue.

★ ★

It's possible to turn lead into gold. However, this process requires a particle accelerator and vast energy. In 1980, American chemist Glenn Seaborg achieved a similar feat by using a particle accelerator to turn bismuth—an element adjacent to lead on the periodic table—into gold. Between 2015 and 2018, scientists at the Large Hadron Collider went a step further and turned lead into gold. However, just 29 trillionths of a gram of gold was produced, lasting for about one microsecond before breaking apart.

Diamonds

Pencil lead and diamond are chemically identical—both substances are composed purely of carbon atoms. The only difference between them is how their carbon atoms are arranged.

★ ★

Diamond is the hardest natural substance on Earth, but it might not be the hardest natural substance on other planets. Scientists believe that under the extreme pressures and temperatures inside some exoplanets, diamond could transform into an even harder, ultra-dense form of carbon known as BC8.

★ ★

Have you ever wondered how diamond can be the hardest natural material on Earth, yet it will still shatter when struck by a hammer? The answer lies in the distinction between 'hardness' and 'toughness'. While 'hardness' refers to a material's resistance to scratching—an area where diamonds excel—'toughness' is a distinct property that refers to a material's ability to withstand breaking—an area where diamonds perform poorly.

★ ★

Approximately 1.5 million microscopic diamonds form in a candle's flame every second.

Water

If all the molecules in a single teaspoon of water were laid end to end, they would stretch for 50 billion kilometres—about eight times the distance from the Sun to Pluto.

★ ★

There are more atoms in one teaspoon of water than teaspoons of water in all the world's oceans.

★ ★

If you slowly pour a handful of salt into a glass filled to the brim with water, the water won't overflow. In fact, the water level will actually drop. This happens because when salt dissolves in water, its ions fit between the water molecules, thereby reducing the overall volume slightly.

★ ★

If you mix 50 millilitres of water with 50 millilitres of ethanol, you will end up with a total of just 96 millilitres. This reduction in volume occurs because the molecules of water and ethanol form hydrogen bonds, allowing them to pack together more compactly.

★ ★

When pure water is cooled below zero degrees Celsius, it doesn't automatically freeze into ice. For pure water to transform into ice, a trigger is needed to initiate the formation of ice crystals. This trigger can be anything, from a speck of dust to a shock wave. Once the trigger is introduced, the crystallisation process begins, and ice forms rapidly throughout the water. To see this phenomenon in action, place a bottle of distilled water into your freezer. The water will cool below freezing point but remain liquid because it's free from impurities. However, if you strike the bottle with your hand, a shock wave will spread through the water, kick-starting the nucleation process—so in an instant, the entire bottle of water will freeze into ice.

★ ★

Pure water does not conduct electricity—it's the dissolved ions and impurities in water that allow electricity to pass through it.

★ ★

Water isn't wet. Scientifically speaking, 'wetness' refers to the interaction between a liquid and a solid surface. This means that water can make other things wet; however, water in its own state is not wet.

★ ★

If you pour just one tablespoon of olive oil into a lake, it will calm the waves across a 2,000-square-metre area. This happens because the oil spreads out to form a one-molecule-thick layer on the water's surface, which prevents the wind from generating waves. During the 19th century, sailors used this technique of dumping oil into the sea to calm violent waves.

★ ★

In 1963, 13-year-old Tanzanian student Erasto Mpemba was making ice cream for his school's cooking class. Short on time, he placed his hot ice cream mixture into the freezer. To his surprise, it froze faster than the cooler ice cream mixtures prepared by his classmates. This led to the discovery of the Mpemba effect, which states that under certain conditions, hot water can actually freeze faster than cold water.

★ ★

When water freezes into ice, it expands. As a result of this phenomenon, an ice cube takes up 9% more volume than the liquid water used to make it.

★ ★

Water is the only non-metallic substance that expands when it freezes. If water didn't possess this unique property, ice would sink rather than float, causing bodies of water to freeze from the bottom up. Such a scenario would make life as we know it on Earth impossible, as organisms wouldn't be able to survive in the liquid water below the surface.

★ ★

Scientists have discovered 20 distinct types of ice, each with its own structure and properties. However, according to computer models, there could be as many as 74,963 different types of ice.

★ ★

Ice is classified as a mineral because it meets all the necessary criteria: it's naturally occurring, solid, and has a crystalline structure.

Food Chemistry

Have you ever wondered why cutting onions makes you cry? When you slice into an onion, it releases enzymes and sulphur-containing compounds, which combine to form a gas called syn-Propanethial-S-oxide. This gas spreads through the air, and when it comes into contact with the moist surface of your eyes, it transforms into sulphuric acid. This creates a stinging sensation, which irritates your eyes and triggers tears.

★ ★

If you rub garlic on the soles of your feet, you will taste it in your mouth within an hour. This happens because allicin—the compound that gives garlic its strong odour—is absorbed through your skin, enters your bloodstream, and travels to your mouth.

★ ★

An egg yolk is so slippery that it's almost impossible to pick up with your fingers. However, if you rub crushed garlic on your fingers, the natural compounds in the garlic, known as mercaptans, create a sticky layer that bonds with the egg yolk, allowing you to lift it easily.

★ ★

Ice cream is simultaneously a solid, a liquid, and a gas. The ice crystals are a solid, the milk and the sugar solution are a liquid, and the air bubbles are a gas.

★ ★

If you place a banana and some green tomatoes together in a paper bag, the tomatoes will turn red within two days. This happens because as bananas ripen, they release ethylene gas—a natural compound that speeds up the ripening process in other fruits.

★ ★

Sodium citrate can transform any cheese into a smooth nacho cheese sauce. Coincidentally, the chemical formula for sodium citrate is Na3C6H5O7.

★ ★

When you taste sour food, you're actually tasting protons.

Pharmacology

Scientists don't fully understand how general anaesthetics work.

★ ★

For 85% of patients, antidepressants are no more effective than a placebo.

★ ★

Botulinum toxin, commonly known as Botox, is primarily used in cosmetic procedures to smooth wrinkles. However, it's also the most poisonous naturally occurring substance in the world. A dose of just 100 nanograms—an amount invisible to the naked eye—is deadly to humans, while one kilogram is enough to kill every person on Earth.

★ ★

Botulinum toxin is one of the most expensive substances in the world, priced at a staggering US$127 trillion per kilogram. To put this cost into perspective, a mere 1.2 grams of the toxin—enough to fill an eyedropper—is equivalent to the value of the International Space Station. The global cosmetic industry uses an annual supply of just a few milligrams of botulinum toxin, which becomes heavily diluted.

★ ★

Early forms of chemotherapy were a by-product of the mustard gas used during World War I.

★ ★

Devrom is an over-the-counter 'internal deodorant' designed to eliminate the unpleasant odours of flatulence and stool at their source. Its active ingredient, bismuth subgallate, binds to the sulphur molecules—the primary cause of these odours—produced by gut bacteria during digestion. This effectively stops embarrassing smells before they occur.

★ ★

Every penicillin strain used in the world today is descended from a single mouldy cantaloupe purchased from an Illinois market in 1943.

★ ★

Antimicrobial resistance is a phenomenon that occurs when bacteria, viruses, parasites, and fungi evolve to resist the drugs designed to treat them. While this resistance can develop naturally, the over-prescription and misuse of antimicrobial medications, such as antibiotics, have significantly accelerated the problem. Every year, an estimated 1.14 million people die globally due to antimicrobial resistance. Without urgent action, the World Health Organisation warns that the global death toll could reach 350 million by 2050.

★ ★

A 2020 study found that it costs an average of US$1.3 billion for a pharmaceutical company to bring a new drug onto the market.

★ ★

The placebo effect is largely a myth. In medical research, studies typically involve two groups: one receiving the medication and the other receiving a placebo. In such studies, those in the placebo group often show improvement. However, this improvement is not due to the placebo effect. Rather, it's actually caused by a statistical phenomenon known as 'regression to the mean', whereby patients naturally tend to improve over time, regardless of intervention. To accurately evaluate the power of the placebo effect, researchers need to include three groups: a treatment group, a placebo group, and a non-treatment group. When such studies have been conducted, the patients in the placebo group and the non-treatment group achieve similar outcomes. This suggests that the placebo effect is largely a myth. Nonetheless, there is one area where placebos might have a slight effect: pain reduction.

★ ★

Producing complex crystalline molecules, which are essential for many medications, is a difficult process on Earth. However, in the microgravity of space, crystals grow more uniformly, form better structures, and attain larger sizes. Because of these benefits, pharmaceutical companies are working to develop drugs in space. Hundreds of drug development experiments have already taken place on the International Space Station. Looking ahead, there are even proposals for a private space station to orbit Earth that's dedicated exclusively to advancing drug research.

Disappearing Polymorphs

A disappearing polymorph is a phenomenon where a crystal structure that was once easy to make suddenly changes into a different crystal structure during nucleation. This change usually occurs because the first type of crystal isn't stable enough, so it naturally shifts to a more stable version. While this new structure has the same chemical makeup as the original, it can have very different physical properties.

★ ★

The disappearing polymorph phenomenon has impacted a variety of pharmaceutical drugs. This has involved just a single microscopic crystal setting off a chain reaction that transforms an entire batch of a drug into a new, less effective version. While it's always theoretically possible to recreate the original drug, doing so is typically impractical because the new version tends to be more stable.

★ ★

In 1996, pharmaceutical company Abbott Laboratories released the HIV drug ritonavir. For the next few years, the drug was produced without any issues. However, in 1998, some ritonavir capsules from the Chicago factory suddenly failed to dissolve properly. Researchers examined the ritonavir under a microscope and found that the drug's crystals had transformed into a new polymorph known as 'form II'. While form II was more stable than standard ritonavir, it was five times less soluble. This made it less effective for patients. It was eventually discovered that the Chicago factory's entire supply of ritonavir had transformed into form II. To investigate why their drug supply had spontaneously changed, a team of scientists from the Chicago factory travelled to the ritonavir factory in Italy. However, they inadvertently carried a few microscopic traces of form II on their bodies. Soon after, the Italian factory's entire ritonavir supply also transformed into form II. This chain reaction continued, so over several weeks, every ritonavir production factory in the world had its supply turn into form II. It was later determined that just a seed of form II—as little as one part per million—would cause all of the ritonavir in an area to transform into form II. Abbott Laboratories eventually resolved the problem by developing a new gelatine capsule formulation of the drug. However, the disappearing polymorph saga cost the company US$250 million.

★ ★

During the 1970s, the antidepressant drug paroxetine anhydrate was patented by the pharmaceutical company Beecham Group.

In 1984, a new hemihydrate version of the drug spontaneously appeared at Beecham Group's UK factories. It was discovered that whenever the original paroxetine anhydrate was exposed to water or humidity, it would transform into the new, more stable hemihydrate form. When Beecham Group's patent for paroxetine anhydrate expired, other drug companies attempted to reproduce the drug. However, any time they tried, it came out as the new hemihydrate version. Scientists believe this occurred because trace amounts of the hemihydrate version had spread into Earth's atmosphere. As a result, whenever the original paroxetine anhydrate was manufactured anywhere in the world, it would immediately interact with the trace hemihydrate crystals present in the atmosphere and transform into the new hemihydrate version. This made the original form impossible to reproduce.

★ ★

A 2018 study found that 15% to 50% of all drugs have a more stable polymorph version waiting to be created. This could potentially render the drugs less effective—or even obsolete.

The Haber-Bosch Process

By the end of the 19th century, agricultural production was struggling to keep pace with the rapidly increasing global population. As a result, scientists around the world were urgently exploring technological advancements to ensure a sufficient food supply for humanity. One critical area of focus was how to provide plants with more nitrogen, a key ingredient in enhancing plant growth. Although nitrogen makes up 78% of the air, it exists in a form that plants cannot use: two atoms of nitrogen triple-bonded together in a difficult-to-break molecule. For many years, scientists attempted to separate this stubborn nitrogen triple bond to unlock its potential for agriculture. Finally, in 1909, German chemist Fritz Haber discovered that under extreme conditions—temperatures of 500 degrees Celsius and pressures equivalent to being two kilometres under the sea—the triple bond could be broken apart. When subject to these conditions, nitrogen molecules break apart and react with hydrogen to form ammonia—a compound that can be used to create nitrogen-rich fertiliser. Building on Haber's discovery, another German chemist, Carl Bosch, refined the process to make it viable on an industrial scale. This innovation became known as the Haber-Bosch process. For more than a century, it has been used to produce fertilisers that have revolutionised the agricultural industry. Thanks to the Haber-Bosch process, farmers can grow four times as much food on the same plot of land, significantly boosting the global food supply. The discovery of the Haber-Bosch process earned Fritz Haber and Carl Bosch the Nobel Prize in chemistry.

★ ★

The Haber-Bosch process is the invention that has saved the most lives throughout history. An estimated four billion people—roughly half the world's population—wouldn't be alive today if the process hadn't been discovered. In fact, there's a good chance you owe your life to the Haber-Bosch process.

★ ★

Roughly half of the nitrogen atoms in your body come from the Haber-Bosch process.

★ ★

Today, the Haber-Bosch process accounts for 1% to 2% of the world's total energy consumption.

★ ★

The Haber-Bosch process has revolutionised food production on a global scale; however, it has also caused significant environmental problems. The highly energy-intensive process releases potent greenhouse gases, contributes to soil degradation, contaminates drinking water, and produces acid rain. Moreover, scientists don't fully understand the long-term environmental effects of converting large quantities of stable, inert nitrogen from the air into highly reactive compounds. In other words, we are in the midst of a worldwide experiment.

★ ★

As the global demand for fertiliser is projected to double over the next century, scientists are actively exploring different ways to reduce the environmental impact of the Haber-Bosch process. These efforts include using renewable energy to power the process, identifying alternative hydrogen sources that don't rely on natural gas or fossil fuels, and employing nanocatalysts to make the process more efficient.

★ ★

German chemist Fritz Haber is often referred to as 'the man who killed millions but saved billions'. His discovery of the Haber-

Bosch process enabled the mass production of fertilisers, which ultimately saved billions of lives. However, he also pioneered the use of chemical warfare by developing chlorine and other toxic gases that caused widespread death and suffering. Additionally, his work on chemical compounds indirectly contributed to the creation of Zyklon B, a pesticide infamously used in Nazi gas chambers to murder over one million Jewish people during the Holocaust.

People and Chemistry

Glenn Seaborg was an American chemist who played a key role in discovering ten elements. Remarkably, he's the only person who's ever been able to write out their address using chemical elements. If Seaborg wanted a letter delivered, he could write Sg, Lr, Bk, Cf, Am. These symbols stand for seaborgium (named after himself), lawrencium (Lawrence Laboratory), berkelium (Berkeley), californium (California), and americium (USA).

★ ★

J.J. Thomson discovered the electron. Thomson's student, Ernest Rutherford, went on to discover the proton. Rutherford's student, James Chadwick, completed the picture by discovering the neutron.

★ ★

In 1915, Polish chemist Jan Czochralski was studying the crystallisation rate of metals when he absent-mindedly dipped his pen into a pot of molten tin, thinking it was an inkwell. When he pulled his pen out, he noticed a long, single crystal hanging from it. Czochralski had accidentally discovered how to produce monocrystalline silicon. Today, Czochralski's technique is the standard for creating monocrystalline silicon and is used in 90% of electronics worldwide that rely on semiconductors.

★ ★

When the Nazis invaded Denmark during World War II, they were on the verge of stealing two Nobel Prize medals made of 23-karat gold. To safeguard these awards, chemist George de Hevesy dissolved them in a potent acidic mixture known as aqua regia. He then poured the resulting liquid into a flask and discreetly stored it on a laboratory shelf in the Niels Bohr Institute, Copenhagen. When the Nazis ransacked the building, they overlooked the seemingly innocuous liquid. Following the war, de Hevesy returned to the laboratory and found the flask untouched. He reversed the chemical process and sent the precipitated gold to the Nobel Society, which used it to recast the medals.

Miscellaneous Chemistry

The field of chemistry has an inconceivably large number of different potential molecular combinations. For example, if you were to create every possible three-dimensional shape using just the chemotherapy drug paclitaxel (chemical formula C47H51NO14), you would fill one-and-a-half observable Universes.

★ ★

Your cells contain an enzyme called OMP decarboxylase, which can accelerate a reaction rate by a factor of 10^{17}. Without this enzyme, a typical reaction would take 78 million years; however, with OMP decarboxylase present, the same reaction takes just 18 milliseconds.

★ ★

The unmistakable 'swimming pool smell' that people associate with chlorine isn't actually produced by chlorine itself—it's the result of a chemical reaction. When chlorine interacts with bodily fluids, like urine or sweat, it forms compounds called chloramines, which create the strong swimming pool odour. This means that if people don't urinate or sweat in the pool, the odour won't be produced. Surprisingly, to get rid of the smell, you actually need to add more chlorine to the pool. The additional chlorine breaks down the chloramines, effectively neutralising the odour.

★ ★

Saltwater itself is odourless. The scent of the sea that you're familiar with is actually a blend of chemicals released by various marine organisms. The sulphury aroma of the sea comes from dimethyl sulphide, a compound released when bacteria eat dead plankton. The fishlike odour of the sea is caused by bromophenols, which fish absorb through their diet of marine worms and algae. The dried seaweed smell of the sea is due to dictyopterenes, the sex pheromones produced by seaweed eggs.

★ ★

Air turns into liquid at around -196 degrees Celsius.

★ ★

Liquid air has a pale blue hue that resembles the colour of water.

★ ★

There's more hydrogen in one litre of petrol than in one litre of pure liquid hydrogen.

★ ★

Avogadro's Law states that equal volumes of any two gases—as long as they are at the same temperature and pressure—will always contain an identical number of molecules.

★ ★

Atoms are constantly being recycled in the environment, circulating through the air, water, and soil. As a result of this recycling, every breath you take likely contains atoms from every breath a human has taken throughout history. For example, Julius Caesar's last gasp in 44 BCE has now spread across the globe. Given the immense number of molecules in a single breath—around 25 sextillion—it's highly probable that the next breath you take will contain at least one molecule from Caesar's last gasp.

★ ★

Your body contains hundreds of billions of atoms that were once in the bodies of Einstein, Hitler, and Cleopatra.

★ ★

A rubber tyre is technically a single, massive, polymerised molecule.

★ ★

In 2004, researchers developed 9,10-Dithioanthracene, the first molecule capable of moving in a straight line by mimicking human walking. The molecule has two linkers that function like feet. One linker remains on the ground, while the other moves forwards, propelling the body of the molecule.

★ ★

The foam in various cleaning products—like shampoo, toothpaste, and soap—doesn't serve a functional purpose. It's only added to the products because consumers associate foam with cleaning.

★ ★

Lead poisoning is responsible for approximately 5.5 million premature deaths worldwide every year—a figure higher than the death toll from HIV, malaria, and car accidents combined. What makes lead particularly dangerous is that there is no safe level of exposure; even microscopic amounts can damage the brain and impact overall health.

★ ★

Lead was first added to fuel in the 1920s to prevent engine knocking. By the 1970s, its harmful effects became widely recognised, leading to its phase-out. However, it wasn't until 2021 that it was eliminated from fuel globally. Removing lead from fuel has had some unexpected benefits. For instance, according to the lead-crime hypothesis, it's responsible for the sharp drop in US crime rates that began in the 1990s.

★ ★

Fluoroantimonic acid is the world's strongest superacid. With a Hammett acidity (pH equivalent) of -28, it's 20 quintillion times more potent than pure sulphuric acid. Because of its extreme corrosiveness, fluoroantimonic acid can only be stored in containers with a Teflon lining. If it's stored in glass, it will dissolve the container.

★ ★

A 2018 study found a link between quasicrystals—unique materials with orderly but non-repeating atomic patterns—and prime numbers. When X-rays were directed through the quasicrystals, researchers observed a regular pattern of bright spots, known as Bragg's peaks. Intriguingly, when the same experiment was repeated, but this time with prime numbers computationally treated as particles, a similar pattern emerged.

★ ★

Glass isn't a liquid or a solid. It's actually an amorphous solid—a unique state between liquid and solid.

★ ★

Wood ants practice chemistry. They gather resin from coniferous trees and combine it with the formic acid they secrete from their abdominal glands. By mixing these two ingredients, they create a powerful antibiotic that shields their colony from disease. This cocktail is the most sophisticated example of pharmacology in the animal kingdom.

COSMOLOGY

The Big Bang

"What happened before the Big Bang?" According to some cosmologists, the answer is simple: nothing happened before the Big Bang. The Big Bang itself was the very beginning of time and space, so there was no 'before' for anything to happen. To illustrate this concept, imagine travelling towards the South Pole. Once you reach it, you can't travel any further south. In the same way, asking what happened before the Big Bang might be meaningless because time and space may not have existed until the Big Bang occurred.

★ ★

Complete the sentence: *"In the beginning, there was a..."*. Many people will answer *"Singularity"*, referring to a moment when all the matter and energy in the Universe was compressed into a single point. However, the singularity model has been outdated for 45 years. Today, cosmologists are almost certain that the Big Bang didn't actually involve a singularity. If the Big Bang singularity model were correct, we would expect to see a variety of clear signatures in the Universe today, but none of these have been observed.

★ ★

The term 'Big Bang' is a misnomer. There was no bomb-like explosion that hurled matter outwards into some pre-existing void. Rather, the Big Bang refers to the expansion of space itself.

★ ★

The Big Bang didn't occur in one specific place. It happened everywhere in the Universe at once, including where you are right now.

★ ★

Cosmologists have a good understanding of what happened in the Universe from about 10^{-43} seconds after the Big Bang onwards. However, our understanding of physics breaks down during the first 10^{-43} seconds after the Big Bang. This is because, at such high energies and tiny scales, Einstein's theory of general relativity conflicts with quantum mechanics. To resolve this discrepancy, physicists are working to develop a new theory called quantum gravity, which aims to combine these two frameworks. A successful theory of quantum gravity would help us understand some of the most profound questions about the beginning of the Universe.

★ ★

At 10^{-33} seconds after the Big Bang, the Universe is believed to have undergone a rapid phase of expansion called cosmic inflation. During this unimaginably brief period—lasting one billionth of a trillionth of a trillionth of a second—the Universe expanded by a factor of 10^{26}. In this fleeting moment, it grew from something much smaller than a proton to about the size of a grapefruit. That's equivalent to a bacterium ballooning up to the size of the Milky Way Galaxy.

★ ★

Cosmologists have virtually no understanding of what the Universe was like prior to inflation. The rapid expansion of the Universe erased any information that would have given us a hint into its origin. All we know is that inflation must have arisen from some pre-existing, non-inflationary state.

★ ★

The Big Bang should have produced equal amounts of matter and antimatter. If this had occurred, all the matter and antimatter would have completely annihilated each other, leaving behind a Universe filled only with energy. However, in the first fraction of a second after the Big Bang, there was a slight imbalance: for every one billion particles of antimatter, there were one billion and one particles of matter. When these particles collided, they annihilated each other, but the one extra particle of matter in every billion survived. This tiny fraction of surviving matter makes up nearly everything we see in the Universe today. Despite numerous theories, the cause of the initial asymmetry between matter and antimatter remains one of the greatest mysteries in physics.

★ ★

If the density of the early Universe had been just 0.00000000001% greater, the Universe would have immediately collapsed back in on itself in a Big Crunch. Conversely, if the density of the early Universe had been just 0.00000000001% lower, the Universe would have expanded so quickly that protons and electrons would never have combined to form atoms.

★ ★

The early Universe was too hot for atoms to form. It took until 380,000 years after the Big Bang for the Universe to cool enough for electrons to combine with nuclei and form the first atoms.

★ ★

Fifteen million years after the Big Bang, the entire Universe was at room temperature. If humans had existed during this period, we wouldn't have needed the Sun to stay warm—the heat from the Big Bang's background radiation would have sufficed.

★ ★

Roughly 1% of the static on an analogue television is actually the background radiation left over from the Big Bang.

Dark Energy

Everything that we can see in the Universe is made up of atoms. However, this visible matter constitutes just 5% of the entire Universe. The remaining 95% is invisible to us and consists of two mysterious components: dark energy, which accounts for 68% of the Universe, and dark matter, which accounts for 27%. The word 'dark' in both terms simply means that their true nature remains unknown.

★ ★

In the late 1990s, it was discovered that the expansion of the Universe is accelerating. To explain this phenomenon, scientists have proposed that the Universe contains an invisible substance with a repulsive gravitational effect. This mysterious force has been named 'dark energy'.

★ ★

The leading theory to explain dark energy is that it's a background vacuum energy permeating all of space. While space is often thought of as empty, it actually contains fluctuating quantum fields that give rise to a background energy. It's theorised that this energy is the dark energy driving the accelerated expansion of the Universe.

★ ★

While vacuum energy is the leading candidate to explain dark energy, it has a major flaw. The amount of vacuum energy that should theoretically exist is a staggering 10^{120} times larger than what has been observed. This enormous discrepancy has been labelled *"the worst theoretical prediction in the history of physics"*.

Dark Matter

According to Newton's laws of motion, stars at the edge of spiral galaxies should orbit significantly more slowly than stars near the centre. However, observations show that stars orbit their galactic centre at roughly the same speed, regardless of their position in the galactic disk. To explain this discrepancy, scientists have proposed the existence of an additional form of matter that provides galaxies with extra gravity. This hypothetical substance has been named dark matter.

★ ★

Dark matter has mass and occupies space, yet it doesn't interact with normal matter in any way, except gravitationally. Furthermore, it doesn't emit, absorb, or reflect any form of electromagnetic radiation, including light, rendering it completely invisible. As a result, scientists have never been able to observe it directly.

★ ★

Dark matter is not just confined to galaxies. It exists in a vast, interconnected web-like structure that permeates the entire Universe.

★ ★

For every kilogram of normal matter in the Universe, there are five kilograms of dark matter.

★ ★

Most cosmologists believe that dark matter is composed of a yet-to-be-identified subatomic particle, such as the weakly interacting massive particle or the axion.

Types of Black Holes

If you compress any object into a small enough volume, it will turn into a black hole.

★ ★

Earth would need to be compressed to a diameter of less than 1.77 centimetres—roughly that of a marble—to become a black hole.

★ ★

Primordial black holes are hypothetical black holes that may have formed in the early Universe, moments after the Big Bang. During this period, the Universe was a hot, turbulent sea of energy with pockets of varying densities. Any region slightly denser than its surroundings may have collapsed under its own gravity into a primordial black hole. If primordial black holes have survived to the present day, they would be the oldest objects in the Universe, even predating atoms.

★ ★

Primordial black holes could theoretically range in size from as small as a subatomic particle up to several hundred kilometres in diameter, yet each would possess an immense mass. For example, a primordial black hole the size of a proton would have a mass equivalent to Mount Everest.

★ ★

Primordial black holes are of great scientific interest because they may explain dark matter, account for mysterious gravitational wave signals, or even serve as the initial seeds for the formation of supermassive black holes.

★ ★

Black holes are categorised into three main types: stellar-mass, intermediate-mass and supermassive. Stellar-mass black holes range from 3 to 60 solar masses, intermediate-mass black holes range from 120 to 100,000 solar masses, and supermassive black holes exceed 100,000 solar masses.

★ ★

There is a gap between 60 and 120 solar masses where no black holes exist. This absence is explained by the pair-instability problem in extremely massive stars. When a star reaches this mass range, it undergoes a catastrophic supernova explosion. This event is so powerful that it completely obliterates the star, leaving no remnant behind and thus no black hole in this mass range.

★ ★

The observable Universe contains an estimated 40 quintillion stellar-mass black holes. These black holes collectively account for about 1% of all normal matter.

★ ★

TON 618 is one of the largest black holes ever discovered. With a radius of 390 billion kilometres—over 40 times the distance from the Sun to Neptune—it's close to the theoretical maximum size that a black hole can reach. Remarkably, its mass exceeds that of all the stars in the Milky Way combined.

★ ★

Supermassive black holes lie at the centre of nearly every large galaxy, yet scientists don't know how they form.

★ ★

Some supermassive black holes formed astonishingly early—within the first few hundred million years after the Big Bang. Scientists remain puzzled over how they could grow to such immense sizes so quickly in the early Universe.

★ ★

Some cosmologists propose that the observable Universe might actually be the interior of an enormous, low-density black hole.

★ ★

One hypothesis proposes that our Universe may have emerged from the turbulent quantum fluctuations that occurred inside a black hole from a pre-existing Universe.

Black Hole Traits

Some supermassive black holes can exist for more than 10^{100} years.

★ ★

Supermassive black holes can be separated by billions of light-years, yet their rotational axes will still be aligned in parallel.

★ ★

Most supermassive black holes have such low densities that they could theoretically float on water.

★ ★

A typical supermassive black hole has a density of about 100 kilograms per cubic metre—equivalent to just 10% the density of water. In comparison, the supermassive black hole located at the centre of the galaxy NGC 1600 is one of the most massive ever observed, yet its density is significantly lower—only one-hundred-thousandth that of water.

★ ★

Virtually none of the stars in a galaxy orbit their central supermassive black hole. While supermassive black holes are extremely heavy, they only constitute a tiny fraction of a galaxy's total mass, so their gravitational influence is limited to a relatively small area. For example, Sagittarius A, the supermassive black hole at the centre of the Milky Way, makes up just one-millionth of our galaxy's total mass; hence, its sphere of influence extends a mere 3.26 light-years and affects only 100 stars. The vast majority of stars in a galaxy orbit the galaxy's overall centre of mass.

★ ★

It's a myth that black holes suck things in like a cosmic vacuum cleaner. In reality, they exert gravitational force based on their mass, just like every other object. In fact, if the Sun were replaced with a black hole of the same mass, all the planets in the Solar System would continue orbiting as they do now.

★ ★

Stable black holes possess just three properties: charge, mass, and spin. However, since a black hole's charge is negligible, knowing its mass and spin is enough to fully describe it.

★ ★

Some black holes spin at 90% the speed of light.

★ ★

Black holes do not actually contain a singularity. A singularity represents a point in the Universe where certain physical qualities become infinite. However, infinite values do not exist in reality. When a theory predicts an infinite value, it indicates that the theory has broken down and is unable to accurately describe extreme conditions in the physical world. Therefore, a new, more robust theory is required.

★ ★

Moments before two black holes collide, they can orbit each other hundreds of times per second.

★ ★

In 2015, scientists detected the gravitational waves produced by the collision of two black holes. At its peak, this event released an astonishing 36 septillion yottawatts of power—50 times more than the energy output from all the stars in the observable Universe combined.

★ ★

In 2024, astronomers discovered the largest pair of black hole jets ever observed. These colossal jets stretch 23 million light-years—a length equivalent to 140 Milky Way galaxies laid end to end.

★ ★

It's theoretically possible to create a black hole using light, due to a phenomenon known as a kugelblitz. To form a black hole with the mass of Earth, you would need to concentrate the light from every star within 350 light-years of our planet, shining continuously for ten years, into a space smaller than a mosquito.

Approaching a Black Hole

Black holes slow down time significantly due to their powerful gravitational pull. To illustrate this phenomenon, imagine a spaceship orbiting the supermassive black hole at the centre of the Milky Way. From Earth's perspective, it would appear as though the spaceship completed one orbit every 16 minutes. However, for the crew on the spaceship, time would pass more slowly due to the black hole's intense gravity, and they would experience one orbit every eight minutes. This means that if the spaceship orbited the black hole for ten Earth years, only five years would pass for the crew on board.

★ ★

If you watched someone approach a black hole, you would never actually see them cross the event horizon. Instead, you'd simply see them stop and slowly fade away.

★ ★

Moments before falling into a black hole, the back of your head would appear in front of you. This bizarre effect would occur because a black hole warps space-time so intensely that the photons reflecting off the back of your head would circle around the black hole and reach your eyes.

★ ★

If you fell into a stellar-mass black hole, the gravitational pull at your feet would be significantly stronger than at your head. This differential in force would stretch your body into a thin, elongated shape, like a piece of spaghetti—a process known as spaghettification.

★ ★

If you crossed the event horizon and fell into a supermassive black hole, you wouldn't notice any unusual gravitational effects at all. This is because a supermassive black hole's immense size creates an extremely gentle gravitational gradient.

★ ★

Inside a black hole, space and time switch places. This means that the centre of a black hole is not a place you can be at—it's a moment you reach.

The Size of the Universe

If the Sun were scaled down to the size of an atom, the observable Universe would be the size of Earth.

★ ★

The Universe is 13.8 billion years old. However, due to the expansion of space, the observable Universe actually has a radius of 46.5 billion light-years.

★ ★

The observable Universe has a diameter of 93 billion light-years. However, according to some models, the entire Universe is more than 250 times as large, extending to at least seven trillion light-years across.

★ ★

Cosmologists don't know whether the Universe is finite or infinite.

★ ★

The Universe must have one of three shapes: open like a saddle, closed like a sphere, or flat like a piece of paper. In an open Universe, two rockets flying side by side would separate and never cross paths again. In a closed Universe, two rockets flying side by side would diverge and eventually meet where they started. In a flat Universe, two rockets flying side by side would always remain parallel. Based on the latest observational evidence, our Universe is most likely flat. However, there's also the possibility that it's curved on an unimaginably vast scale. If this is the case, the curvature would be so subtle that, within our 93 billion light-year view of the Universe, we simply cannot see a large enough portion of it to detect the curvature.

★ ★

If we had spacecraft capable of travelling at the speed of light, 94% of the galaxies in the observable Universe would already be permanently beyond our reach. This is because the continuous expansion of the Universe is causing the space between unbounded objects to increase. This expansion of space is so rapid that galaxies beyond a distance of 14.5 billion light-years are being pushed away faster than light can travel. As a result, it's already too late for humans to ever reach 94% of the observable Universe.

★ ★

Space is constantly expanding, causing galaxies to drift farther apart over time. As a result of this expansion, in the distant future, any intelligent civilisation gazing into the cosmos will only be able to see their own galaxy. All the other galaxies will have drifted so far away that they will be beyond the limits of observation. This means that, for such a civilisation, their own galaxy will effectively become their entire observable Universe.

★ ★

We don't know how fast the Universe is expanding. Cosmologists have used two different methods to determine the rate of the Universe's expansion, but the results don't align. The first method, which involves measuring tiny fluctuations in the cosmic microwave background radiation, suggests the Universe is expanding at 67 kilometres per second per megaparsec. The second method, which involves measuring pulsating stars known as Cepheid variables, suggests the Universe is expanding at a faster rate of 74 kilometres per second per megaparsec. This discrepancy, known as the Hubble tension, remains one of the greatest mysteries in cosmology.

★ ★

Imagine a long, narrow cylinder extending across the 93 billion light-year diameter of the observable Universe. If the circular face of this cylinder were equal to the cross-sectional area of a uranium nucleus, the entire cylinder would have a volume of just 150 litres.

The Multiverse

The multiverse theory proposes that our Universe, with its hundreds of billions of galaxies, is just one amongst countless other universes. Many of the world's leading cosmologists subscribe to a variant of this theory.

★ ★

One of the most popular multiverse theories is the eternal inflation model. Just 10^{-33} seconds after the Big Bang, the Universe is believed to have undergone a rapid expansion known as inflation. This inflationary period was incredibly brief, lasting just 10^{-32} seconds. The eternal inflation model suggests that when inflation ends in one region, it continues in others. This means that when inflation ceased in our Universe, it may have persisted elsewhere, creating countless other expanding Universes. Each of these Universes would have its own unique set of particles, laws of physics, forces, and values of fundamental constants. Within this vast multiverse, some universes would, by chance, possess the finely-tuned properties necessary to support life.

★ ★

A prominent multiverse theory is the many-worlds interpretation of quantum mechanics. This theory offers a solution to an issue raised by the famous Schrodinger's cat thought experiment. In this experiment, a cat is placed in a box with a radioactive substance. If even a single atom of the substance decays, it triggers a Geiger counter, releasing poison and killing the cat. However, according to quantum mechanics, the atoms in a radioactive substance exist in multiple states simultaneously—they only commit to a specific state when observed. This implies that until the box is opened, the cat is theoretically both dead and alive. However, in the real world, a cat cannot physically be both dead and alive—and this is a major problem in physics. The many-worlds interpretation resolves this paradox by proposing that every time a quantum system interacts with a larger object, the universe splits into multiple branches. In these branches, every possible outcome of the quantum event occurs. In the case of Schrodinger's cat, the universe divides into two: in one branch, the cat survives, and in the other, it dies. This branching process happens continuously, creating an infinite number of parallel universes, where each possible outcome of every quantum event is realised.

★ ★

According to some versions of string theory, our Universe may exist as a three-dimensional brane—akin to a 3D island—floating in higher-dimensional space. This brane would be just one of many possible branes, each potentially housing its own universe existing parallel to ours. These other parallel universes would be separated from us by dimensions that we cannot directly perceive.

★ ★

The latest estimate from string theory suggests there could be $10^{272,000}$ universes.

★ ★

Cosmologist Max Tegmark has proposed a controversial, yet intriguing, multiverse theory known as the Mathematical Universe Hypothesis. According to Tegmark, every piece of mathematics actually exists and corresponds to the real world. For example, imaginary numbers explain the behaviour of quantum particles, and non-Euclidean geometry describes gravity. However, there is a vast amount of mathematics, and only a small fraction of it manifests in our physical world. Tegmark believes that the mathematical structures that don't show up in our universe appear in the real worlds of other universes.

★ ★

Some versions of the multiverse theory propose that you have an infinite number of doppelgangers.

★ ★

Our Universe could one day collide with another Universe. According to certain multiverse theories, in an infinite multiverse, new 'bubbles' might continuously form, each containing a separate universe. If one of these bubbles were to form near the edge of our bubble, a collision would occur, and we would expect to see evidence of it. In fact, such a collision may have already taken place. In 2015, cosmologist Dr. Ranga-Ram Chary mapped the cosmic microwave background—the light leftover from the Big Bang—and detected a mysterious glow 4,500 times brighter than expected. One proposed explanation for this glow is that it represents matter leaking into our Universe from another universe.

The Simulation Hypothesis

The simulation hypothesis, devised by philosopher Nick Bostrom, proposes that our Universe could be an advanced digital simulation, akin to a highly sophisticated computer game. According to this idea, a civilisation with immense computational power would have the potential to run detailed simulations on its futuristic computers. These simulations could be so complex and realistic that the beings within them would believe they live in a real, physical universe, completely unaware it's just a simulation. Bostrom argues that an advanced civilisation could run a vast number of simulations, and each simulated society could, in turn, create its own simulations. Therefore, it becomes statistically more likely that we are the virtual beings created by an advanced civilisation, rather than the original, non-simulated beings.

★ ★

The Universe exhibits a variety of characteristics that support the simulation hypothesis. First, it operates entirely through algorithms and mathematical equations, which suggests that our reality can be fundamentally reduced to information processing. If you were a character in a computer game, you would eventually discover these laws of nature, which are essentially the programming code. Second, the Planck length—the smallest measurable unit of space—might be analogous to the pixels in a computer game, representing the resolution of our reality. Similarly, the speed of light—the maximum speed limit in the Universe—may reflect the processing capacity of the system running the simulation. Finally, the Universe's remarkably fine-tuned constants and conditions support the notion that it may have been intentionally designed by an advanced intelligence.

★ ★

Astronomer David Kipping has analysed the simulation hypothesis using Bayesian probability. His findings suggest there is a 50%

chance that we are living in a simulation. Kipping believes that if humans were ever to create a simulation featuring conscious beings, then it would be almost certain that we ourselves also lived in a virtual world.

★ ★

There are a number of arguments against the simulation hypothesis. First, the energy, data, and processing power required to simulate an entire universe in real-time would be staggering, even for an advanced civilisation. Second, we have no evidence that consciousness can be simulated. It's possible that genuine consciousness is not something that technology can produce, regardless of sophistication. Third, if an advanced civilisation is simulating us, then who created them? Are they also a simulation? This leads to an infinite regression, so eventually there must be a base reality that is not simulated—it's simpler to say that we are that base reality. Fourth, the simulation hypothesis is not falsifiable, meaning it may fall outside the realm of science and be closer to a metaphysical belief. Finally, according to Occam's razor, the simulation hypothesis adds an enormous layer of unnecessary complexity. It requires an undiscovered base reality, an advanced civilisation of creators, and unfathomable computational technology. The simplest explanation is that our Universe is real and operates according to the laws of physics that we are familiar with.

★ ★

Scientists have proposed a variety of ways to test the simulation hypothesis. One approach is to search for 'glitches' or unusual patterns that might indicate limitations in the Universe's resolution or processing power. These could appear as irregularities in high-energy particle collisions or tiny inconsistencies in the precision of physical constants. Another approach is to examine the behaviour of particles at the smallest of scales. If the Universe is simulated, it might be built on a grid-like framework that imposes constraints on how particles move and interact.

Boltzmann Brains

The Boltzmann brain thought experiment states that, given enough time, random particle fluctuations in the void of space will spontaneously form a fully functioning human brain. This brain might even be equipped with an entire lifetime of fabricated memories. In fact, you could be a Boltzmann brain at this very moment, unaware that your existence and memories are not real.

★ ★

Do you think it's more likely that random particle fluctuations would create a Boltzmann brain or an entire universe that eventually gives rise to human brains? Given that a Boltzmann brain is far smaller and simpler than an entire universe, it's statistically much more likely to be formed.

★ ★

In an eternally expanding universe, Boltzmann brains would eventually far outnumber all the ordinary human brains that could ever exist. It follows that, statistically speaking, you are most likely a Boltzmann brain.

★ ★

It would take roughly 10^10^50 years of random quantum fluctuations for a Boltzmann brain to spontaneously form in a vacuum and persist long enough to have a single coherent thought.

★ ★

Most cosmologists do not believe that you are actually a Boltzmann brain. However, the concept of Boltzmann brains does highlight flaws in our current understanding of the evolution of the Universe. As physicist Brian Greene has noted, *"I am confident that I am not a Boltzmann brain. However, we want our theories to similarly concur that we are not Boltzmann brains, but so far it has proved surprisingly difficult for them to do so"*.

False Vacuum Decay

The Universe could be destroyed at any moment, without warning, by a phenomenon called false vacuum decay. While most quantum fields are completely stable, the Higgs field—which gives particles their mass—might exist in a state of precarious instability known as a 'false vacuum'. If the Higgs field were to collapse at even a single tiny point in space, it would create a bubble that expands infinitely in all directions at the speed of light. This bubble would obliterate everything in its path, including all life.

★ ★

The best way to assess the likelihood of false vacuum decay occurring is by measuring the mass of the Higgs boson particle. A lighter Higgs boson would indicate that the Universe is in a stable state, whereas a heavier Higgs boson would suggest that the Universe is in a precarious state of instability. Scientists have recently measured the mass of the Higgs boson and found that it lies near the boundary of 'stable' and 'unstable', meaning that the Universe's stability is uncertain.

★ ★

If false vacuum decay were to occur, it would change the very nature of physics as we know it. The physics described in textbooks would become obsolete, replaced by an entirely different framework. In this new reality, fundamental particles would behave differently, atoms would be held together in new ways, and chemicals would react unpredictably. Ultimately, this new physics would render life in the Universe impossible.

★ ★

If false vacuum decay has already begun somewhere in the Universe, we wouldn't know about it until it reached Earth and wiped us out. However, given the immense scale of the Universe, it could take billions of years to get here. Moreover, if the decay started sufficiently far away, it would never reach us. This is because even though the decay expands at the speed of light in all directions, the ongoing expansion of the Universe would prevent it from ever arriving.

The Fate of the Universe

As of 2025, most cosmologists believe that the Universe will expand forever, in a scenario known as the Big Freeze. Stars will continue to form for the next 100 trillion years, until they exhaust their fuel and fade away. The Universe will then grow darker, becoming dominated by black holes for an almost incomprehensible span of time—approximately 10^{106} years. To put this timeline into perspective, it's equivalent to the Universe having just one second of stars, followed by one billion billion billion billion billion billion billion years of only black holes. We live in that one second of brightness.

★ ★

According to physicist Roger Penrose, once all the black holes have evaporated, the Universe will be essentially empty, containing little more than photons. At that point, the entire mass of the Universe will have transformed into energy, meaning there won't be any way to keep time. As the arrow of time will have vanished, it will be impossible to distinguish moving forwards in time from moving backwards in time. The Universe will also have lost its ability to define scales of space, meaning that the distinction between big and small will become meaningless.

★ ★

Due to random quantum tunnelling and fluctuations, in approximately 10^10^10^56 years, our Universe could potentially generate a new Big Bang, leading to the birth of entirely new universes.

★ ★

Cosmologists are uncertain whether dark energy maintains a constant strength or changes over time. If dark energy is growing stronger, it could lead to a catastrophic phenomenon known as the Big Rip. This would involve dark energy becoming a runaway phantom energy that tears the Universe apart billions of years from now. If this were to occur, we would first notice clusters of galaxies drifting apart. Our night sky would then darken as stars receded from the Milky Way. Three months before the end, the Earth would separate from the Sun, followed by the Moon detaching from the Earth. In the final hours, the Earth itself would explode. At the last instant, molecules, atoms, and even atomic nuclei would disintegrate. Finally, everything in the Universe, including space-time itself, would be ripped apart.

★ ★

Data collected by the Dark Energy Spectroscopic Instrument in 2024 and 2025 suggests that dark energy reached its peak strength 4.5 billion years ago and is now about 10% weaker. If this weakening trend continues, the Universe could ultimately end in a scenario known as the 'Big Crunch'. According to this model, the Universe's expansion would slow, halt, and then reverse, so the Universe would collapse in on itself under the influence of gravity. The Universe would become a giant fireball, and at the very end, neither time nor space would remain. One variation of the Big Crunch is the Big Bounce theory. This idea proposes that the collapse of the Universe would trigger a new Big Bang, giving birth to a fresh Universe. According to this cyclic model, the Universe might be locked in an endless loop of expansions and contractions, seesawing between Big Bangs and Big Crunches for eternity.

Miscellaneous Cosmology

Einstein's theory of general relativity predicts the existence of wormholes—hypothetical tunnels connecting distant regions of space. In theory, wormholes could provide a shortcut for space travellers; however, their practical use appears highly unlikely because they would collapse almost instantly. To keep a wormhole open would require a hypothetical substance called exotic matter, which has negative mass.

★ ★

According to string theory, everything in the Universe is composed of incredibly small, vibrating strings. To illustrate the scale of these strings, if an atom were enlarged to the size of the observable Universe, a single string would be the size of a tree.

★ ★

The Poincare recurrence theorem states that in any self-contained, finite system, every possible state will eventually repeat itself, given enough time. Applying this theorem to the Universe—assuming it's finite, isolated, and deterministic—it has been calculated that in 10^10^10^10^2.08 years, all the particles in the Universe will return to the exact positions they occupy at this very moment. In other words, in the extremely distant future, the Universe will appear exactly as it does now. However, practically speaking, this timescale is so astronomically long that a new phase of the Universe will likely occur before the Universe repeats itself. The Poincare recurrence timescale for the Universe—10^10^10^10^2.08 years—is one of the largest numbers in physics. It's so incomprehensibly immense that its units are irrelevant. 'Years' can be replaced by Planck time (10^{-43} seconds), hours, or even millennia—yet the number remains the same.

★ ★

One of the most fundamental principles in physics is the law of conservation of energy, which states that energy cannot be created or destroyed—it can only change forms. While this principle holds true in our everyday experiences, in the expanding Universe, energy isn't conserved.

★ ★

The likelihood that our Universe's initial low-entropy state arose purely by chance was astonishingly low. In fact, physicist Roger Penrose has calculated that the odds were approximately one in 10^10^123.

★ ★

Have you ever wondered why we live in a Universe with precisely three spatial dimensions? According to one theory, at the moment of the Big Bang, the Universe actually had nine spatial dimensions; however, only three of these dimensions expanded.

★ ★

If the Universe had three dimensions of time and one dimension of space—instead of the other way around—the speed of light would be the minimum speed limit in the Universe, rather than the maximum speed limit. In such a Universe, everything would be composed of tachyons—theoretical particles that always travel faster than light.

★ ★

The odds of finding an exact copy of yourself in the observable Universe are incredibly low: about one in 10^10^68. However, if the Universe is infinite—due to the limited number of ways that matter can be arranged—it's estimated that an identical copy of you is currently living approximately 10^10^28 metres away.

EARTH SCIENCE

Earth's Magnetic Field

Earth's magnetic north and south poles are misnamed. Earth's magnetic north pole is actually the south pole of a magnet, and vice versa. This is why the north end of a compass magnet points north: it's being attracted to what is really the magnetic south pole.

★ ★

Earth's magnetic poles are constantly shifting due to the turbulent movement of liquid iron and nickel in the planet's outer core.

★ ★

From 1900 to 1980, Earth's magnetic north pole remained relatively stationary near the Canadian Arctic. However, in 1980, it began moving rapidly and is now drifting towards Siberia at 60 kilometres per year.

★ ★

Earth's magnetic south pole is currently located outside the Antarctic Circle.

★ ★

The South Atlantic Anomaly is a region over the South Atlantic Ocean where Earth's magnetic field is significantly weaker than the rest of the planet. Satellites and spacecraft passing through this area are exposed to increased radiation levels, which can damage their electronic systems. To minimise this exposure, satellite operators often temporarily power down sensitive instruments and non-essential systems while traversing this area.

★ ★

During the Roman era, Earth's magnetic field was twice as strong as it is today.

★ ★

Over the past 200 years, Earth's magnetic field has weakened by 9%.

★ ★

Some scientists believe the recent weakening of Earth's magnetic field may indicate that a geomagnetic reversal is imminent—an event in which Earth's magnetic north and south poles switch places. Over the past 83 million years, there have been 183 such reversals—an average of one every 450,000 years. These reversals can occur as frequently as every 10,000 years or as rarely as every 50 million years. The most recent successful geomagnetic reversal took place 780,000 years ago.

★ ★

Earth's next geomagnetic reversal will take place gradually over thousands of years. During this process, the planet's magnetic field will weaken significantly, potentially diminishing to one-tenth of its normal strength. This will leave Earth more vulnerable to harmful solar and cosmic rays. In the short term, this radiation will severely disrupt orbiting satellites, including those responsible for controlling electrical grids, potentially causing decade-long blackouts. In the long term, some scientists speculate that the heightened radiation exposure from a geomagnetic reversal could render parts of the planet uninhabitable, leading to mass extinctions.

Earth's Rotation

Across billions of years, Earth's rotation has gradually slowed, causing the length of a day to increase. When the Moon formed 4.5 billion years ago, a day on Earth lasted just ten hours. By 1.4 billion years ago, a day had increased to 18 hours and 41 minutes. During the dinosaur era, a day had lengthened further to 23 hours. Currently, a day on Earth lasts 23 hours, 56 minutes, and 4 seconds. In 200 million years, a day is expected to increase to 25 hours.

★ ★

Without the Moon, a day on Earth would last just six to eight hours, and a year would consist of 1,100 to 1,400 days.

★ ★

Scientists can determine the length of a day on Earth hundreds of millions of years ago by studying fossilised coral. Similar to tree rings, coral produces a thin layer of calcium carbonate each day it grows. By counting these daily layers between seasonal growth rings, scientists can estimate how many days made up a year in the past. For example, coral from 430 million years ago displays 420 daily layers between seasonal rings, indicating that a year during this period consisted of 420 days.

★ ★

Due to a combination of Earth spinning faster at the equator and time dilation, people in Longyearbyen, Norway, experience one extra second every 25,000 years compared to those living in Ecuador.

★ ★

The Coriolis effect is the apparent deflection of moving objects due to Earth's rotation. As the planet spins, objects moving over long distances appear to curve: rightwards in the Northern Hemisphere and leftwards in the Southern Hemisphere. The Coriolis effect influences ocean currents, weather patterns, and even the trajectory of projectiles. However, contrary to popular belief, the Coriolis effect is too weak to influence small-scale, short-duration movements, like the direction of water swirling down a drain or toilet.

★ ★

When military snipers aim at long-range targets, they must take into account the Coriolis effect. This is because at distances exceeding one kilometre, Earth's rotation can cause a bullet to deviate by several centimetres—enough to miss the target.

★ ★

In the Northern Hemisphere, rivers typically erode their right bank slightly more than their left bank, while in the Southern Hemisphere, the opposite occurs. This phenomenon is caused by the Coriolis effect.

★ ★

North-south running railway tracks often have one side visibly wearing out faster than the other. This is because the Coriolis effect causes trains to exert slightly more pressure on one side of the tracks as they move.

Earth's Interior

In 1970, the Soviet Union launched the Kola Superdeep Borehole project—an ambitious attempt to drill as deep as possible into Earth's crust. After two decades of drilling, the borehole reached a depth of 12,262 metres, making it the deepest human-made hole in history. This depth even surpassed that of the Mariana Trench, the deepest point in Earth's oceans. However, the project was permanently halted in 1992, when temperatures of 180 degrees Celsius were encountered, rendering further drilling impossible. Despite its record-breaking depth, the Kola Superdeep Borehole extends just 0.19% of the distance to the centre of the Earth. To put this depth into perspective, if the Earth were the size of an apple, the borehole would not have come close to penetrating the apple's skin.

★ ★

Approximately 2,000 kilometres beneath Earth's surface lie enormous structures known as large low shear velocity provinces (LLSVPs). These massive formations of hot mantle stretch as wide as continents and tower up to 100 times the height of Mount Everest. Yet, despite their immense size, scientists remain puzzled by their nature, origin, and purpose. A leading theory suggests that LLSVPs may actually be the leftover material from Theia—an ancient planet that likely collided with the early Earth, resulting in the formation of the Moon.

★ ★

Earth's inner core is as hot as the surface of the Sun.

★ ★

Nearly all of Earth's internal heat comes from just two sources: the leftover heat from Earth's formation and the radioactive decay of certain elements.

★ ★

All the gold that has ever been mined would fill just three Olympic-sized swimming pools.

★ ★

More than 99% of the gold on Earth is located in the core. In fact, according to some models, there is so much gold in the core that it could cover Earth's entire land surface to a depth of your knees.

★ ★

Due to gravitational time dilation, Earth's core is approximately 2.5 years younger than its surface.

The Atmosphere

Believe it or not, Earth's atmosphere actually extends beyond the Moon. The outermost region of Earth's atmosphere, known as the geocorona, consists of a cloud of hydrogen atoms that stretches up to 630,000 kilometres from Earth's surface—about 1.5 times the distance to the Moon.

★ ★

Earth has experienced three distinct atmospheres throughout its history. The first atmosphere consisted of the hydrogen and helium present in the early Solar System. The second atmosphere was made up of gases released by ancient volcanoes. The third, and current, atmosphere developed 2.5 billion years ago, when cyanobacteria began producing oxygen through photosynthesis.

★ ★

Between 2005 and 2019, the amount of heat trapped in Earth's atmosphere doubled.

★ ★

Scientists expect the ozone layer to fully heal within the next 40 years.

★ ★

If you look in the opposite direction to the Sun just before sunrise or just after sunset, you'll notice a blue band adjacent to the horizon. This is Earth's shadow being projected onto the atmosphere.

★ ★

Scientists have detected mysterious, ultra-low-frequency sounds in the stratosphere, 21 kilometres above Earth's surface. These sounds occur several times every hour, yet they haven't been traced to any known source.

★ ★

Every rainbow is unique for the observer. When multiple people see the same rainbow, each person views the water droplets from a slightly different angle, resulting in subtle variations in the colours they perceive.

★ ★

A rainbow always appears at an angle of 40 to 42 degrees from the shadow of your head.

★ ★

There are 12 distinct types of rainbows, including one that is entirely red.

★ ★

A moonbow is a rainbow that occurs at night. It forms when moonlight refracts through water droplets in the atmosphere, resulting in a fainter version of the traditional solar rainbow.

★ ★

A rainbow's true shape is a full circle. However, we typically only see a semicircle because the Earth's surface blocks the lower half of the rainbow. When viewed from the air, a rainbow appears as a full circle.

Lightning

A lightning bolt can reach a temperature five times hotter than the surface of the Sun.

★ ★

A typical lightning bolt is several kilometres long, yet has a width of just 2.5 centimetres.

★ ★

A single lightning bolt can release a peak power of one terawatt—enough to power the entire United States for the brief duration of the bolt's existence.

★ ★

A 'bolt from the blue' is a lightning bolt that travels horizontally up to 40 kilometres from a storm cloud, before striking the ground. These bolts are particularly dangerous because they can appear to come out of a clear blue sky, catching people off guard.

★ ★

Cosmic rays from the Sun and supernovas can help trigger lightning by electrifying the air.

★ ★

Lightning can start on the ground and travel upwards to a cloud.

★ ★

Lightning can be classified as either negative or positive, depending on the type of charge transferred. Negative lightning makes up 95% of all lightning strikes, while positive lightning accounts for the remaining 5%. Despite being less common, positive lightning is significantly more powerful and dangerous than negative lightning. It's also responsible for the majority of forest fires and powerline damage.

★ ★

A superbolt is an extremely rare and powerful form of lightning that occurs just once in every 250,000 lightning strikes. This intense type of bolt is up to 1,000 times brighter than regular lightning and can release up to 1,000 times as much energy. Interestingly, superbolts were first recorded in 1977 by satellites from the Vela program that were designed to detect clandestine nuclear weapons explosions.

★ ★

Dark lightning occurs when a thunderstorm's powerful electric field emits a burst of gamma rays. If a person is struck by dark lightning—such as while flying through a thunderstorm—they can receive a radiation dose equivalent to a full-body CT scan.

★ ★

Dark lightning can shoot beams of antimatter into space.

★ ★

Ball lightning is a mysterious phenomenon, typically associated with thunderstorms, that's characterised by the appearance of a glowing, floating orb. This orb can range in size from as small as a pea to several metres in diameter. It can move unpredictably or hover above the ground, lasting anywhere from a few seconds to several minutes. Despite extensive research, scientists don't understand the precise cause of ball lightning. In fact, a 2010 study theorised that half of all ball lightning sightings may actually be hallucinations caused by the intense magnetic fields present during severe storms.

★ ★

Never take a shower or use running water during a storm. If lightning strikes your house, it can send an electrical charge through your metal plumbing, potentially electrifying your tap water. In the US alone, an estimated 10 to 20 people are injured every year from using household water during storms.

★ ★

For every one-degree Celsius rise in the global average air temperature, there is a 12% increase in lightning strikes.

★ ★

Lightning strikes occur twice as frequently above busy shipping lanes. This is because ship exhausts release aerosol particles into the atmosphere that contribute to the formation of ice crystals in clouds. These ice crystals increase the clouds' electrical charge, resulting in a higher frequency of lightning strikes.

The Water Cycle

Half of the world's annual precipitation falls on just 12 days of the year.

★ ★

Most precipitation begins as snow. As it falls to Earth, it melts and turns into rain.

★ ★

A water molecule spends an average of nine days in the atmosphere before falling to Earth.

★ ★

Once a water molecule lands in the ocean, it takes an average of 3,000 years before it evaporates again.

★ ★

The Moon influences rainfall. When the Moon is directly overhead or underfoot, it increases the air pressure, resulting in slightly less rainfall.

★ ★

If all the water in the atmosphere were to rain down at once, Earth's surface would be covered to a depth of just 2.5 centimetres.

★ ★

Contrary to popular belief, falling raindrops are not teardrop-shaped. They are actually rounded on the top and flat on the bottom—the same shape as the top half of a hamburger bun.

★ ★

An estimated 40% of all falling rainfall evaporates mid-air, before it ever reaches the ground. This phenomenon can sometimes be observed in the form of wispy, jellyfish-like clouds, known as virga, that appear to hang down from the sky.

★ ★

Despite covering 71% of Earth's surface, the origin of our planet's water—when and how it arrived—remains a mystery.

★ ★

Approximately 232 million years ago, during the Carnian pluvial episode, Earth experienced an intense period of heavy rainfall that lasted one to two million years.

★ ★

The average cumulus cloud contains 500 tonnes of water. That's equivalent to the weight of 100 elephants suspended above your head.

★ ★

At any given moment, clouds cover about two-thirds of the Earth's surface.

The Ocean

When people picture the ocean, they typically imagine crystal-clear waters teeming with life. Yet, in reality, about 94% of the ocean's volume is shrouded in permanent darkness because sunlight cannot penetrate more than 200 metres below the surface. Without sunlight, photosynthesis is impossible, leaving the vast ocean depths devoid of plant life and home to only a sparse population of animals. These hardy creatures survive on alternative energy sources, such as sinking organic matter or the chemical-rich plumes of hydrothermal vents.

★ ★

If you were to cut yourself ten metres below the ocean's surface, your blood would appear green. That's because water absorbs red light, leaving the green wavelengths visible. This same principle explains why so many deep-sea creatures are red. In the ocean's depths, there is no red light to reflect, so red creatures appear black and virtually invisible to predators.

★ ★

The pressure in the Mariana Trench, the deepest point in the ocean, is so extreme that it's equivalent to one person trying to hold up 50 jumbo jets.

★ ★

The top three metres of the ocean's surface contain as much heat as Earth's entire atmosphere. This is due to water's significant mass and high heat capacity compared to air.

★ ★

Over the past 25 years, half of the observed rise in sea levels has been caused by the thermal expansion of water—when water heats up, it expands.

★ ★

If every ice sheet and glacier on Earth were to melt, sea levels would rise by 70 metres.

★ ★

Approximately 230 million people worldwide live in areas less than one metre above sea level at high tide.

★ ★

Earth's oceans contain roughly 18 million tonnes of gold, which, if extracted, would be worth approximately US$1 quadrillion—ten times the current global GDP. However, this gold is dispersed in tiny particle form at an extremely low concentration of about one part per billion.

★ ★

Oceanic waves can travel for tens of thousands of kilometres when not impeded by land. For example, waves originating near New Zealand can reach Alaska, while some waves near Antarctica can circle the continent indefinitely.

★ ★

Amphidromic points are unique spots in Earth's oceans where there are no tides. They occur due to the interaction between the shape of the oceanic basin, the depth of the water, and the Coriolis effect. Rather than experiencing the usual rise and fall of tides, the water around these points rotates in a circular pattern. Amphidromic points can be found in various locations, including the North Sea, the Gulf of Mexico, the Pacific Ocean, and even in many places bordering the coastline.

Volcanoes

Approximately 80% of all volcanic eruptions on Earth occur in the ocean.

★ ★

Volcanic activity was a major factor in four of Earth's five mass extinctions.

★ ★

The Toba supervolcano, which erupted in Indonesia 74,000 years ago, nearly wiped out humanity. The eruption was so immense that it buried parts of Malaysia under nine metres of ash and dimmed the Sun for six years. According to the Toba catastrophe theory, the event may have even plunged the planet into a 1,000-year cooling episode and drastically reduced the human population, leaving behind as few as 3,000 surviving individuals to rebuild humanity.

★ ★

The 1991 eruption of Mount Pinatubo in the Philippines released 17 million tonnes of sulphur dioxide into the atmosphere, causing global temperatures to drop by 0.5 degrees Celsius for the next two years.

★ ★

Roughly 10% of the world's population live within 100 kilometres of an active volcano.

★ ★

Over 80% of Earth's surface is volcanic in origin.

Earthquakes

When people feel an earthquake, they often post about it almost instantly on social media. In fact, by the time an earthquake's seismic waves travel 100 kilometres from the epicentre, they're already being overtaken by social media posts about the event. This means that people more than 100 kilometres from the epicentre often learn about an earthquake through social media before experiencing the tremors for themselves.

★ ★

The Richter scale has not been used since the 1970s. Today, most seismologists measure earthquakes using the moment magnitude scale.

★ ★

Earthquakes are measured using a logarithmic scale, where each whole-number increase in magnitude represents a 32-fold increase in energy release. For example, a magnitude 7.0 earthquake releases 32 times more energy than a magnitude 6.0 earthquake and 1,000 times more energy than a magnitude 5.0 earthquake.

★ ★

The magnitude of an earthquake is limited by the length of the tectonic fault on which it occurs. This means that an earthquake of magnitude 10 or higher is impossible on Earth because no tectonic fault is long enough to generate one. Such an event would require a fault spanning nearly the entire planet. The most powerful earthquake in recorded history struck off the coast of Chile in 1960, registering a magnitude of 9.5.

★ ★

Large earthquakes can trigger other large earthquakes on the opposite side of the globe, within 30 degrees of the antipode.

★ ★

The magnitude 9.0 earthquake that struck Japan in 2011 generated S-waves that travelled through the Earth's interior and triggered 1.5-metre-high waves in Norway just 30 minutes later.

★ ★

Deep earthquakes can sometimes be felt by people on the other side of the planet. For example, the magnitude 8.3 earthquake that struck 600 kilometres beneath Japan's Kuril Islands in 2013 was felt by people as far away as Moscow and Australia.

★ ★

Earthquakes can have negative magnitudes. For instance, a cat falling off a table would register as the equivalent of a magnitude -2 earthquake, a feather floating to the ground would register as the equivalent of a magnitude -7 earthquake, and a single dust particle landing on a table would register as the equivalent of a magnitude -15 earthquake.

Natural Disasters—Other

The winds generated by an average cyclone produce 1.5 trillion watts of power. That's equivalent to one-sixth of the world's total electrical generating capacity.

★ ★

A hypercane is a theoretical type of cyclone that could form if sea surface temperatures rise to 50 degrees Celsius. Such a cyclone could last for several weeks and produce winds near the speed of sound. Some scientists speculate that the Chicxulub impactor, which struck Earth 66 million years ago, may have triggered a series of hypercanes that contributed to the extinction of the dinosaurs.

★ ★

In deep water, a tsunami races across the ocean at around 700 kilometres per hour—roughly the speed of a commercial jet. However, as it nears the coast, it slows down considerably.

★ ★

People often think of tsunamis as towering waves; however, in the open ocean, they are usually less than a metre high—so small that ships typically don't notice them passing. As they approach the shore, they grow dramatically because the seafloor rises, compressing the wave energy into a narrow water column.

★ ★

A rogue wave is an unexpected wave that is at least twice the height of the surrounding waves. Unlike tsunamis, which are caused by geological events, rogue waves are produced by powerful currents, strong winds, and wave interactions. For many years, the scientific community dismissed rogue waves as merely a sailor's myth. However, this perception changed in 1995, when a 25.6-metre-high rogue wave was recorded by scientific instruments off the coast of Norway.

★ ★

A forest fire can lie dormant beneath layers of vegetation or snow for months or even years on end. Then, when the conditions become warm and dry, it can re-emerge and spread again.

★ ★

A fire moves faster uphill than downhill. In fact, its speed doubles for every ten-degree increase in slope.

Geology

Kyawthuite is the rarest mineral on Earth—only one specimen has ever been discovered. Its extreme rarity stems from the fact that it can only form under extraordinarily unique geological conditions:

intense pressure, high temperatures, and a precise concentration of antimony and bismuth in a specific oxidising environment.

★ ★

Itacolumite is a naturally occurring type of rock that becomes bendable when cut into thin strips.

★ ★

The Great Unconformity is a puzzling gap in Earth's geological record, where layers of rock that should have formed over hundreds of millions of years are mysteriously missing. Normally, sediment builds up gradually, creating a continuous timeline of Earth's history. Yet, these layers are conspicuously absent from 1.2 billion to 250 million years ago, leaving older rocks abruptly layered beneath much younger ones. Geologists don't know what caused these rock layers to be missing. Possible explanations include massive erosion from ancient glaciers, tectonic shifts, or major sea-level fluctuations. The Great Unconformity is one of the most baffling mysteries in geology.

★ ★

Radiocarbon dating is unreliable for fossils over 50,000 years old. To determine the age of such fossils, scientists date the rock layers above and below them.

★ ★

The oldest Earth rock ever discovered was actually found on the Moon.

★ ★

Under the right conditions, pyrite crystals can naturally form into an almost perfect cube shape. This occurs when the ideal combination of pressure, minerals, and moisture causes the iron and sulphur atoms to bond into a precise lattice pattern.

★ ★

Mexico's Giant Crystal Cave is home to the largest crystals on Earth, with some reaching up to 11 metres in length and 1.2 metres in width—about the size of mature pine trees. These crystals, which grow by just one millimetre every 2,200 years, take millions of years to reach their impressive sizes.

★ ★

Piezoelectricity is a property exhibited by certain crystals that allows them to generate an electrical charge when mechanical pressure is applied. It's used in many everyday technologies, such as quartz clocks for accurate timekeeping, microphones for converting sound into electrical signals, inkjet printers for precise ink droplet control, and ultrasound machines for medical imaging.

★ ★

It took until the late 1960s for the geoscience community to accept the theory of plate tectonics.

★ ★

In roughly 250 million years, all the continents on Earth will likely merge into a single supercontinent.

★ ★

Marine fossils can be found atop the world's tallest mountains. This is because mountaintops were once the bottom of ancient oceans. Over millions of years, tectonic plates collided and pushed the seafloor upwards, transforming it into towering mountains.

★ ★

Mountains have roots that extend deep beneath Earth's surface, much like the submerged part of an iceberg. These roots are usually five times deeper than the mountain's visible height above ground. This means that if a mountain is 5,000 metres tall, its roots might extend 25,000 metres below the surface.

★ ★

When a mountain peak erodes, you might expect the mountain to shrink in height. Yet, surprisingly, the opposite occurs—the mountain actually grows taller. This happens due to a process known as isostasy, whereby the erosion of material from the peak means there's less mass pushing down on the mountain, causing it to float higher on Earth's crust. Due to isostasy, a 5,000-metre-high mountain would need to lose more than 25,000 metres of material through erosion before it begins to shrink in height.

★ ★

The most abundant material on Earth is not water, sand, or soil—it's something that most people have never even heard of: bridgmanite. This magnesium-iron silicate, which is buried deep within the planet's interior, makes up a staggering 38% of Earth's volume. Despite its prevalence within Earth, it was first identified in a meteorite that landed in Australia in 1879 and was only officially named in 2014.

Fossil Fuels

Most of the coal we use today comes from plants that lived 200 to 400 million years ago, during the Carboniferous period. Over millions of years, these plants were buried under layers of earth, where heat and pressure gradually transformed them into coal.

★ ★

The majority of the oil we use today comes from ancient marine microorganisms—like plankton and algae—that lived in prehistoric oceans 100 to 300 million years ago. Over time, their remains were buried beneath layers of sediment, where heat and pressure slowly transformed them into oil.

★ ★

Termite flatulence is responsible for approximately 2% of the world's methane emissions.

★ ★

In the 13th century, Genghis Khan inadvertently cooled the planet by killing 40 million people—roughly 11% of the world's population. With fewer people alive, there was a reduction in farming, industry, and wood burning, resulting in 700 million tonnes of carbon dioxide being removed from the atmosphere.

★ ★

The air pollution caused by burning fossil fuels is responsible for one in five deaths globally.

★ ★

Every year, from January to March, a 6,000-kilometre-long layer of pollution called the Asian brown cloud settles over South Asia. This cloud is made up of smoke, dust, and other harmful chemicals. It blocks the sunlight, affects weather patterns, contributes to serious health problems, and causes the deaths of hundreds of thousands of people annually.

★ ★

In the UK and Australia, more people die from motor vehicle pollution than from traffic accidents.

★ ★

You are more likely to die from exposure to plane exhaust pollutants than in a plane crash.

★ ★

Cycling can be bad for the environment. A 2006 paper by Karl Ulrich found that each year of cycling adds 10.6 days to a person's life. However, due to this increase in lifespan, cyclists end up consuming more fossil fuels over their lifetime than non-cyclists, potentially making cycling worse for the environment than driving a fossil-fuelled car. If a cyclist wishes to truly reduce their overall fossil fuel footprint while increasing their lifespan, they need to modify their diet, electricity usage, and shopping habits.

Microplastics

Microplastics are tiny pieces of plastic measuring less than five millimetres in diameter. Alarmingly, they are now found in nearly everything we eat and drink. In fact, scientists estimate that the average person unknowingly ingests around five grams of microplastics every week—equivalent to the weight of a credit card. The consumption of microplastics has been linked to various health conditions, including immune system disorders, endocrine disruption, heart disease, and cancer. Studying the health effects of microplastics is challenging for researchers because the particles are present in everyone, so there's no control group for comparison. Although microplastics are ubiquitous in our food system, you can reduce your exposure by avoiding bottled water, teabags, plastic cutting boards, and foods packaged in plastic.

★ ★

A 2024 study found that, by weight, the human brain is composed of 99.5% brain and 0.5% microplastics.

★ ★

Between 2016 and 2024, the amount of microplastics in the human brain increased by 50%.

★ ★

Microplastics have been discovered in the placentas of unborn babies, raising concerns about their potential impact on fetal development.

★ ★

One litre of bottled water contains an average of 240,000 pieces of microplastics.

★ ★

A 2024 study found that drinking bottled water is linked to several health risks, including a 5% increased risk of hypertension, a 9% increased risk of diabetes, a 21% increased risk of stomach ulcers, and a 17% increased risk of kidney stones. The study authors speculate that the microplastics in bottled water may be responsible for these increased health risks.

★ ★

You should never put plastic in the microwave. A 2023 study found that microwaving plastic containers—even those labelled 'microwave-safe'—can release billions of microplastics into food. This occurs because heat accelerates the degradation of plastic, breaking it down into smaller fragments that can easily leach into your meal. To minimise your exposure to microplastics, experts recommend using glass or ceramic containers when heating your food in the microwave.

★ ★

Plastic is already pervasive in the air and soil at alarming levels, and by 2050, it's estimated that the weight of plastic in the ocean will likely exceed the weight of all the fish. If we continue using plastic at current rates, it's likely to lead to an ecological disaster on a global scale.

Anthropology

All humans are remarkably similar on a genetic level—in fact, two chimpanzees from different regions of Africa are more genetically diverse than all of humanity. The lack of diversity in modern humans is primarily the result of population bottlenecks in our relatively recent past.

★ ★

Approximately 1.2 million years ago, the effective global early human population was estimated to be around 18,500 individuals. For the next million years, the population stayed at such critically low levels that early humans would have been classified as an endangered species.

★ ★

Around 74,000 years ago, the human population plummeted to just a few thousand individuals. Every person alive today is descended from this small group of survivors.

★ ★

Just 300,000 years ago, our Homo sapiens ancestors lived alongside eight other distinct species of humans.

★ ★

Neanderthals are named after the Neander Valley in Germany, where their first recognised fossils were discovered in 1856. By pure coincidence, the word 'Neander' derives from a German surname meaning *"new man"*.

★ ★

If Earth's history were condensed into a 24-hour period, modern humans would appear just one second before midnight.

★ ★

Of the 117 billion people who have ever lived, 27% died before their first birthday.

★ ★

The last day all humans were on Earth together was 30 October 2000. Since that date, there have always been people aboard the International Space Station.

★ ★

Earth is currently going through a sixth mass extinction event, called the Holocene extinction, which has been driven by human activity. Studies indicate that extinction rates for plants and animals are now 100 to 1,000 times the natural background rate—and only accelerating. Alarmingly, if this trend continues, some scientists project that half of Earth's complex lifeforms could be extinct by 2100.

★ ★

Using conservative estimates, Oxford philosopher Toby Ord has calculated there is a one-in-six chance of human extinction occurring within the next century. He identifies the most pressing threats as uncontrollable artificial intelligence and an engineered pandemic.

Earth and Space

A 2024 study found that Earth likely had a Saturn-like ring system around 466 million years ago. Researchers believe these rings cast a shadow on Earth's surface that may have contributed to a dramatic global cooling period called the Hirnantian Icehouse.

★ ★

It's well known that the gravitational pull of the Sun and the Moon create ocean tides. However, they also generate terrestrial tides, causing Earth's solid land surface to rise and fall by up to 55 centimetres per day.

★ ★

The COVID-19 pandemic cooled the Moon. During April and May 2020, roughly half of the world's population was under some form of lockdown, resulting in a sharp decline in greenhouse gas emissions and aerosols. This meant that Earth's surface trapped and re-emitted less heat, so less heat reached the Moon. As a result, the Moon's nighttime temperature fell by eight to ten degrees Celsius.

★ ★

If you could drive your car vertically, it would take about one hour to reach space.

★ ★

If Earth were scaled down to the size of a basketball, outer space would be just two millimetres above the ball's surface.

★ ★

When you watch a sunset, the Sun is already completely below the geometric horizon. It only appears above the horizon due to a phenomenon known as atmospheric refraction, whereby Earth's atmosphere bends the Sun's light, making it visible even after the Sun has technically set.

The Chicxulub Impactor

Approximately 66 million years ago, an asteroid known as the Chicxulub impactor struck Mexico's Yucatan Peninsula, triggering the extinction of the non-avian dinosaurs.

★ ★

The Chicxulub impactor is often depicted in movies as a slow-moving object that gradually grows larger in the sky as it approaches Earth. But in reality, it was travelling 20 times faster than a speeding bullet—covering the distance from an aeroplane's cruising altitude to the ground in a mere 0.3 seconds. In other words, it went from being invisible to striking the Earth in the blink of an eye.

★ ★

The Chicxulub impactor punched a hole in our atmosphere, creating a vacuum leading directly into outer space. This hole persisted for a few seconds before the atmospheric pressure equalised and the surrounding air rushed back in to fill the gap.

★ ★

Remarkably, the Chicxulub impactor began forming a crater even before it hit the ground. This is due to the sheer force of its shock wave and its immense energy.

★ ★

The Chicxulub impactor was so massive—10 to 15 kilometres in diameter—that at the time of impact, its peak was higher than present-day Mount Everest.

★ ★

When the Chicxulub impactor collided with Earth, it released one billion times as much energy as the atomic bombs dropped on Nagasaki and Hiroshima.

★ ★

The Chicxulub impactor was so powerful that it raised part of Earth's crust 25 kilometres high. This left a towering rim of mountains around the crater's edge that were higher than the present-day Himalayas.

★ ★

The Chicxulub impactor triggered a massive tsunami, with waves that initially soared up to 4.5 kilometres high. The fossilised megaripples from this ancient tsunami can still be seen in Louisiana today.

★ ★

Debris from the Chicxulub impactor rained down on the opposite side of the planet, sparking widespread forest fires that engulfed 70% of the world's forests.

★ ★

Some of the molten material ejected by the Chicxulub impactor reached temperatures several times hotter than the surface of the Sun.

★ ★

The Chicxulub impactor ejected an astonishing 25 trillion tonnes of material into the atmosphere—enough to fill the Baltic Sea. This debris formed a dense cloud of dust that blocked out the Sun and caused a global winter lasting 15 years.

★ ★

There are probably fragments of dinosaur bone on the Moon. The Chicxulub impactor was so powerful that it launched 70 billion kilograms of rock into space. It's likely that some of this debris, which would have included dinosaur remains, ended up on the Moon.

★ ★

A 2013 study found that the Chicxulub impactor propelled rocks—and possibly even dinosaur fragments—as far as Mars and the moons of Jupiter.

★ ★

Some scientists believe that the Chicxulub impactor catapulted microscopic Earth organisms onto Jupiter's moons, where they could still be alive today.

★ ★

The Chicxulub impactor dispersed a layer of iridium—a rare element commonly found in asteroids—right across the globe. This iridium is still visible around the world today in the 66-million-year-old layer of rock known as the K-Pg boundary.

★ ★

Scientists still haven't determined the specific 1,000-year window in which the Chicxulub impactor struck Earth. However, they have established that the collision occurred during the Northern Hemisphere's springtime. They reached this conclusion by studying ancient fish fossils that contained traces of molten and vaporised rock from the Chicxulub impactor. Because fish bone growth is seasonal, the scientists were able to establish that the fish died during the springtime.

★ ★

The Chicxulub impactor wiped out 75% of all species.

★ ★

The Chicxulub impactor caused Earth's climate to warm for 100,000 years.

★ ★

If the Chicxulub impactor had struck Earth just 30 seconds earlier or later, the planet's rotation would have caused it to land in a safer area. If this had occurred, dinosaurs would probably still roam Earth today, and humans would never have evolved.

★ ★

In 1980, it was first proposed that an asteroid impact may have caused the extinction of the dinosaurs. However, it took until 2010 for the international scientific community to formally accept this idea.

Miscellaneous Earth Science

The Earth pulsates every 26 seconds, yet scientists don't know why.

★ ★

Around 2% of the global population can hear a persistent, low-frequency sound known as the 'worldwide hum'. Scientists don't know the source of this sound, or even whether it's caused by a single phenomenon.

★ ★

Every decade, from the 1960s to the 2010s, the average global terrestrial wind speed decreased by 0.5 kilometres per hour. It's not known what caused this decline.

★ ★

The wind itself is silent. What we perceive as 'the sound of the wind' is actually the wind passing through objects, brushing against surfaces, or vibrating against our ears.

★ ★

The strength of gravity on Earth's surface varies by approximately 0.7%, depending on the location. Gravity ranges from 9.7639 m/s^2 in the Peruvian Andes to 9.8337 m/s^2 on the surface of the Arctic Ocean. The commonly cited value of 9.8 m/s^2 represents the average gravitational strength across Earth's surface. The variation in Earth's surface gravity is caused by several factors, including the planet's non-spherical shape, uneven density distribution, and the centrifugal force at the equator.

★ ★

If Earth were 50% larger in diameter, gravity would be so strong that we wouldn't be able to venture into space using rockets.

★ ★

If Earth's diameter were doubled, the increase in surface gravity would cause most trees and plants to collapse under their own weight.

★ ★

If you cut the Earth into one-metre cubes, then lined them up end to end, they would stretch across the diameter of the Milky Way. Similarly, if you cut the Earth into one-millimetre cubes, then lined them up end to end, they would stretch across the diameter of the observable Universe.

★ ★

Around 650 million years ago, during the Cryogenian period, Earth experienced a dramatic event known as 'Snowball Earth', where the entire planet was covered in snow and ice for millions of years. This extreme glaciation made Earth resemble a giant, frozen snowball.

★ ★

Technically, Earth is still in an ice age. An ice age is defined as a prolonged period during which the planet has extensive ice sheets and glaciers, especially at the poles and at higher latitudes. The current ice age is known as the Quaternary Ice Age. It began approximately 2.6 million years ago and continues to this day.

★ ★

The Borsuk-Ulam theorem states that there will always be two points on exactly opposite sides of the Earth with the same temperature and air pressure, down to the infinitesimal fraction.

★ ★

Oklo, Gabon, is the only known place on Earth where natural nuclear fission reactors once existed.

★ ★

The boiling point of water varies with altitude. At sea level, water boils at 100 degrees Celsius. However, for every 152.4-metre increase in elevation, the boiling point drops by 0.5 degrees. As a result, at the summit of Mount Everest, water boils at just 70 degrees Celsius.

HUMAN BODY

Perception Limits

The human fingertip is so sensitive that it can detect bumps as small as 13 nanometres in height—roughly the thickness of 24 stacked silicon atoms. To put this into perspective, if your finger were scaled up to the size of the Earth, you could feel the difference between cars and houses.

★ ★

The human cheek is so sensitive that it can feel something as light as a fly's wing being dropped on it from a height of 7.6 centimetres.

★ ★

On a clear night, the human eye can see a candle being lit from 2.8 kilometres away.

★ ★

The human eye can detect a single photon of light.

★ ★

The human eye can distinguish between ten million different shades of colour.

★ ★

The human eye can naturally see objects as small as 0.1 millimetres in length—such as a human egg cell.

★ ★

The human nose can detect geosmin—the chemical compound that gives fresh rain its scent—at concentrations as low as five parts per trillion. This level of sensitivity surpasses a shark's ability to smell blood in water.

★ ★

The human nose can smell ozone at concentrations as low as ten parts per billion.

★ ★

The human nose can detect a single drop of perfume diffused through a three-room apartment.

★ ★

The human nose can distinguish between more than one trillion different odours.

★ ★

The human ear can hear the difference between hot and cold water being poured.

★ ★

The human ear can detect a wristwatch ticking from six metres away.

★ ★

Humans can taste the toxic alkaloid strychnine, even when diluted to a concentration as low as 1.4 parts per million.

The Mouth

If an adult tooth gets knocked out, try to place it back into its socket as soon as possible. In most cases, the periodontal ligament will reattach the tooth to the socket. (Note: After placing the tooth back into its socket, you still need to be examined by your dentist).

★ ★

A toddler's upper row of adult teeth is located just below their eyes. As the child grows, these teeth gradually move downwards, eventually replacing the baby teeth.

★ ★

US company Lantern Bioworks has developed a genetically modified bacterium that may permanently cure tooth decay. With just one brushing, the bacterium will outcompete and replace the harmful bacteria in your mouth that cause cavities, potentially providing you with lifelong protection against tooth decay.

★ ★

In 2024, Japanese researchers began trials of the world's first tooth regrowth drug. The drug works by blocking a protein in the mouth that normally prevents natural tooth regeneration, thereby allowing new teeth to grow automatically. Successful animal trials have already been conducted, meaning the drug could be commercially available as soon as 2030.

★ ★

Nearly half of all US adults over the age of 30 have gum disease.

★ ★

The use of cutlery has altered the structure of the human mouth. Humans historically had an edge-to-edge bite; however, the adoption of cutlery 250 years ago allowed us to cut our food into smaller, more manageable pieces. This change in eating habits modified our jaw mechanics, leading to the development of an overbite.

★ ★

Taste receptors are not just found in the mouth—they're located right throughout the body. These extraoral taste receptors exist in organs like the stomach, pancreas, lungs, brain, and even the skin. They detect chemical compounds in the environment and trigger essential responses. For example, bitter taste receptors in the airways can sense harmful substances and activate protective reflexes, while sweet receptors in the stomach help manage glucose absorption.

★ ★

You can't taste food without saliva—your taste buds can only detect chemical molecules once they're dissolved in liquid. This can be demonstrated with a simple experiment. Gently dry a small part of your tongue using a clean paper towel. Next, place a pinch of a dry substance, like salt or sugar, onto that dry spot. Interestingly, you won't be able to taste it at first. However, if you wait a few moments, your salivary glands will naturally moisten the area, dissolving the substance, allowing you to taste it again.

★ ★

The taste of food is not only influenced by its flavour but also by its smell, touch, and appearance. This can be easily demonstrated through a few simple experiments. First, eat a jelly bean while pinching your nose shut—you'll notice that it loses much of its flavour. Second, leave a biscuit out of its packaging overnight—it will turn soggy and lose its taste. Finally, add food colouring to lemonade and ask someone to drink it—they'll struggle to identify the beverage.

★ ★

When someone's eyes are closed and their nose is pinched shut, they typically can't taste the difference between a potato peel and an apple peel.

★ ★

Have you ever wondered why mint creates such a cool, refreshing sensation in your mouth? The answer lies in menthol, a natural compound found in mint. Menthol binds to the TRPM8 receptor—a protein in your body that detects cold temperatures. When activated, this receptor sends the same 'cold' signal to your brain that an ice cube does.

★ ★

Contrary to popular belief, your tongue doesn't naturally rest on the floor of your mouth—it actually rests on the roof. This position, known as 'tongue posture', plays a key role in supporting proper breathing, swallowing, and overall oral health.

★ ★

It's actually possible to tickle yourself! To do so, gently rotate your tongue in a circular motion on the roof of your mouth.

★ ★

Moments before you vomit, your mouth produces extra saliva that is slightly alkaline. This saliva acts as a protective buffer, shielding your mouth and teeth from the harsh stomach acids in the forthcoming vomit.

The Nose

At any given time, you do about 75% of your breathing through one nostril and 25% through the other. Every few hours, your dominant nostril switches. To identify your dominant nostril, stand close to a mirror and exhale through your nose. You'll see two fog marks on the mirror—one for each nostril. The larger fog mark corresponds to your dominant nostril.

★ ★

You have a stereo sense of smell. When you smell something, your brain pinpoints the direction of the scent by comparing the tiny difference in the number of odour molecules reaching each nostril. If one nostril is blocked, this comparison isn't possible, making it harder to locate the source of a smell.

★ ★

Olfactory fatigue is a phenomenon whereby prolonged exposure to a specific smell reduces your ability to detect it. It occurs in many everyday situations—like entering a restaurant and initially noticing the aroma of the food, only to become less aware of it over time; applying perfume or aftershave and at first sensing its strong scent, but gradually growing accustomed to it; and not being able to smell your own breath because you are so used to the odour. To overcome this last example of olfactory fatigue and smell your own breath, lick your wrist and wait about ten seconds. The scent of your breath on your skin will be much more noticeable.

★ ★

When a person shakes hands with someone of the same gender, they often unconsciously sniff their own hand afterwards. This behaviour is believed to be a way of picking up chemical signals left by the other person, which might provide subconscious information about them. Interestingly, this instinctive act is similar to how dogs sniff each other when they meet.

★ ★

Joy Milne is a Scottish woman with hereditary hyperosmia, a rare condition that grants her an extraordinary sense of smell. In a 2019 experiment, she sniffed the shirts of 12 people—six with Parkinson's and six without—to see whether she could detect those with the disease. Milne correctly identified 11 of the 12 cases and was convinced that one person, although not diagnosed with Parkinson's at the time, actually had the condition. Eight months later, that individual was indeed diagnosed with Parkinson's.

★ ★

It's impossible to hum while holding your nose closed. Humming requires air to flow from your voice box out through your nostrils, but when your nose is blocked, this airflow is disrupted.

Ears

When you hear a faint sound, your eardrums move less than one picometre—a distance 100 times smaller than the diameter of a hydrogen atom.

★ ★

If you were to draw a dot on your eardrum using a permanent marker, that dot would appear on the outer surface of your ear within a few weeks. This happens because the skin lining your ear canal grows outwards in a spiral pattern to help remove earwax and dead skin cells. (Note: You should never attempt this—poking the eardrum is dangerous).

★ ★

Nausea triggered by benign paroxysmal positional vertigo can be cured by performing a series of head movements that help reposition the displaced crystals in the inner ear.

★ ★

Right-ear advantage is a phenomenon whereby most people can hear speech better with their right ear than their left. It occurs because speech heard by the right ear travels directly to the brain's left hemisphere, where language is processed. In contrast, speech heard by the left ear first travels to the brain's right hemisphere, then crosses to the left hemisphere through a pathway called the corpus callosum. This detour causes a delay of between 10 and 20 milliseconds, making speech perception slower and less accurate. Right-ear advantage is particularly noticeable in children. For example, a typical seven-year-old can accurately repeat information heard through their right ear 70% of the time, and their left ear just 55% of the time.

★ ★

Most people can hear music better with their left ear than their right. This is because the left ear is more directly connected to the right side of the brain, which is responsible for musical processing.

★ ★

Infrasound is inaudible to humans, yet it can still impact us by causing a feeling of awe, unease, or even the sensation that a supernatural event has occurred. Some researchers speculate that this phenomenon may explain why people often perceive old buildings as being haunted—wind passes through the building's cracks and loose fittings, generating infrasound.

★ ★

As people age, the hair cells in their cochlea that detect high-frequency sounds gradually deteriorate. Because of this decline, you can estimate someone's approximate age by giving them one of the many high-frequency hearing tests found online.

★ ★

Your ear circumference increases by about 0.51 millimetres per year.

Eyes

Just as you are left or right-handed, you probably also have a dominant eye. Your brain prefers this eye because it provides more input for tasks that require focus and precision. For most people, eye dominance corresponds with handedness, so if you are right-handed, you are likely right-eye dominant. To determine your dominant eye, extend your arms in front of you and form a triangular shape by touching your thumbs and index fingers. With both eyes open, look through this triangular opening at a distant object, such as a doorknob. Now, close your left eye. If the distant object remains centred in the triangle, you are right-eye dominant. If you close your right eye and the object remains centred, you are left-eye dominant.

★ ★

Each of your eyes has a blind spot. This is a region where there are no light receptors because the optic nerve and blood vessels exit the eye. Your brain uses the surrounding visual context to 'fill in' your blind spot with what it expects to be there. To locate your blind spot, take a piece of paper and draw a dot about ten centimetres to the left of a plus sign. Hold the paper 50 centimetres away from your face, close your right eye, and focus on the plus sign. Move the paper slowly towards you. When the dot disappears, you've found the blind spot in your left eye. To test your right eye, rotate the paper and repeat the process.

★ ★

At night, your peripheral vision becomes sharper than your central vision. This happens because your rods, the photoreceptor cells responsible for low-light vision, are concentrated around the outer edges of your retina.

★ ★

As the sun sets, your eyes begin a process known as dark adaptation, whereby they gradually become more sensitive to low levels of light. This adjustment typically takes about 20 to 30 minutes—roughly the same time it takes to get dark at the equator after sunset. Through dark adaptation, your eyes can detect light sources across an astonishing brightness range—spanning a factor of one billion.

★ ★

Most people are trichromats, meaning they have three types of cones in their eyes that enable colour vision. However, a small percentage of the population—primarily women—are tetrachromats, who possess a fourth type of cone. This additional cone allows some tetrachromats to perceive 100 times as many colours as the rest of the population.

★ ★

If you notice someone with red eyes in a photograph, you're actually seeing the blood at the back of their eyeballs. When a camera takes a photograph, it emits a quick burst of light that catches the eyes off guard, leaving the pupils no time to contract. As a result, the light travels deep into the eyes, illuminating the blood-rich retinas—making the eyes appear red in the photograph.

★ ★

If a person has just one red eye in a photograph, it can be a sign of a serious eye condition, such as eye cancer.

★ ★

The transparent outer layer of the eye, known as the cornea, is one of the few tissues in the human body without a blood supply. It absorbs its oxygen directly from the surrounding air.

★ ★

Just 10,000 years ago, all humans had brown eyes.

★ ★

People with aphakia, a condition in which the lens of the eye is missing, can see ultraviolet light.

★ ★

Tim Doucette is a Canadian man who was born blind and underwent surgery to remove his lenses. As a result of this surgery, he gained the ability to see faint, deep-sky objects in space that are typically invisible to the naked eye.

Visual Phenomena

When you gaze up at a clear blue sky, you'll notice tiny white dots moving across your field of vision. These dots are actually the white blood cells flowing through your eyes. They follow the paths of your blood vessels and briefly accelerate with every heartbeat. This visual effect is known as the blue field entoptic phenomenon.

★ ★

You can see the blood vessels in your own eye, thanks to the Purkinje tree phenomenon. To observe this effect, either poke a pin through a card to create a small hole or curl your index finger to form a tiny opening. Next, find a bright white background, like a blank document on a computer screen. Close one eye and look through the hole, while shaking it in a five-millimetre-wide circular motion. As you do this, a network of grey lines will appear on the background, resembling a spider's web. These are the blood vessels in your eye.

★ ★

You literally 'see' the world upside down. When you look at your surroundings, the convex lenses in your eyes project inverted images onto your retinas. Your brain then processes these two flipped images and combines them into a single, right-side-up image of the world. During the seventeenth century, German priest Christoph Scheiner demonstrated this process by replacing a bull's retina with a screen. He noticed that the image of the scene in front of the bull's eye appeared inverted on the screen. This proved that the image that forms on the retina is upside down, before the brain corrects it.

★ ★

You can demonstrate that images are cast onto your retina upside down by performing a simple experiment. You'll need two index cards and a pin. Start by taking the first index card and using the pin to poke three tiny holes, two millimetres apart, arranged in the shape of a small triangle. On the second index card, poke a single hole. Next, hold the card with the three holes close to your eye, and hold the card with the single hole around 12 centimetres farther away. Face a gentle light source, such as a window, and look through the three holes towards the single hole. You should see an inverted triangle. This proves that the image is flipped when it reaches your retina.

★ ★

Try scanning your surroundings from left to right with smooth eye movement. You'll quickly realise this is impossible—your eyes don't move smoothly but instead jump from point to point in rapid movements called saccades. During these jumps, your brain blocks your vision, making you momentarily blind. However, you don't see black during these jumps. Instead, you experience a phenomenon called saccadic masking, whereby your brain fills in the gap by showing you either the last image before the jump or the next stable image. This prevents you from perceiving motion blur during eye movement. Remarkably, due to saccadic masking, you are effectively blind for about 40 minutes every day.

★ ★

To experience saccadic masking firsthand, try this simple experiment. Look at your reflection in a mirror, and while keeping your head perfectly still, focus on your left eye. Now, quickly switch your gaze to your right eye. Because of saccadic masking, you won't actually be able to see your eyes move between positions. Repeat the experiment, but this time while recording yourself with your phone's front-facing camera. When you play back the video, you'll see your eyes moving—something your brain typically hides from your perception.

★ ★

Have you ever glanced at an analogue clock and noticed that the second hand appears stationary for more than one second? This effect, called the stopped-clock illusion, occurs due to saccadic masking. When your eyes move to focus on the clock, your brain momentarily blocks your vision during the movement, and fills in the gap with the first stable image it sees—the second hand of the clock in its new position. This makes it appear as though the second hand has paused for longer than one second.

★ ★

Just one to two degrees of your visual field—roughly an area the size of your thumbnail at arm's length—is in high resolution. The rest of your visual field is in low resolution, so it's out of focus. Every second, your eyes make three to four rapid movements, known as saccades, which provide your brain with enough input to create a sharp image of your environment.

★ ★

To understand just how narrow your high-resolution vision is, focus on a single letter in this book. You'll likely only recognise about two to three letters on either side of that point.

★ ★

You live in a three-dimensional world, yet each of your eyes sees it as flat. This occurs because your eyes are positioned apart on your face, so each one captures a slightly different two-dimensional image of your environment—a phenomenon called binocular disparity. Your brain then compares these two images, analyses the slight differences between them, and uses this information to calculate depth. This is why you perceive the world in three dimensions.

★ ★

Binocular disparity can be demonstrated with a simple experiment. Start by holding your hands about 20 centimetres in front of your face at eye level. Next, point your index fingers towards each other, leaving a one-centimetre gap between them. Now, look through this gap at a distant object, such as a wall. You should see what appears to be an extra floating finger. This occurs because while you're focused on the distant object, each of your eyes provides your brain with a different image of your close-up fingers. When your brain overlaps these images, they don't line up, so your brain creates the illusion of the extra floating finger.

★ ★

Your nose is always within your field of vision, so your eyes can technically see it at all times. However, your brain uses a process called unconscious selective attention to filter out unnecessary details, effectively making your nose disappear from your perception. To temporarily bypass this effect and see your own nose, simply close one eye.

★ ★

The McCollough effect is a visual phenomenon that occurs when a person stares at red horizontal and green vertical stripes for several minutes. After this exposure, if the person looks at black and white stripes, they will perceive them as tinted with red and green. What makes the McCollough effect particularly remarkable is its persistence. The illusion can last for days, weeks, or even months. In rare instances, the effect has been reported to become permanent.

★ ★

The Himba tribe of Namibia doesn't have a specific word for the colour blue. As a result, when shown one blue square and 11 green squares, they have difficulty distinguishing the blue from the green. However, the Himba language has more words for shades of green than English does. Therefore, when presented with 12 green squares—one of which is a slightly different shade—they can instantly spot the odd one out, a task that the rest of the world finds challenging.

★ ★

Unlike English, Russian doesn't have a single word for 'blue'. Instead, it uses distinct terms for light blue (goluboy) and dark blue (siniy). Thanks to this linguistic distinction, Russian speakers can differentiate between shades of light blue and dark blue significantly faster than English speakers.

Blindness

A person with total blindness doesn't see blackness or darkness—they experience no visual sensation whatsoever. It's the equivalent of trying to see out of your elbow.

★ ★

To experience what it feels like to 'see' nothing, try this experiment. Close one eye and focus on an object directly in front of you with your open eye. While your open eye sees the object, your closed eye provides no visual information to your brain, so it 'sees' nothing at all.

★ ★

In blind individuals, the primary visual cortex—the region of the brain that typically processes visual information—is rewired, so it instead responds to sound. This adaptation allows many blind people to understand speech at a rate of up to 25 syllables per second. That's three times faster than what sighted individuals can comprehend.

★ ★

During the 1970s, neuropsychologist Lawrence Weiskrantz conducted a series of experiments on people who had total vision loss caused by cortical blindness. In one experiment, Weiskrantz showed the subjects a series of shapes—circle, square, or triangle—and asked them to identify which shape they had been shown. Despite reporting that they couldn't see, the subjects guessed the correct shape at a rate significantly better than chance. In a case study, a man with total cortical blindness was asked to walk through a cluttered corridor. Remarkably, he managed to navigate the space without colliding with a single object. In another case study, a man with total cortical blindness was shown objects on a screen. He was able to identify them with over 90% accuracy, even though he had no conscious awareness of seeing them. Through these experiments, Weiskrantz discovered the phenomenon of blindsight, in which individuals with total cortical blindness can respond to visual stimuli despite having no conscious perception of sight. Cortical blindness occurs when the primary visual cortex—the brain region responsible for processing vision—is damaged. People with this condition still receive light in their eyes and convert it into neural information; however, the damage to their primary visual cortex prevents this information from reaching conscious awareness. Instead, other parts of their brain take over and process the visual data. This allows them to accurately guess what they are 'seeing' without having a conscious awareness of it.

★ ★

There has never been a documented case of a person who was born blind developing schizophrenia.

★ ★

In 1688, Irish polymath William Molyneux wrote a letter to the philosopher John Locke. In it, he posed the question: Imagine a person, blind from birth, who has learned to distinguish between a cube and a globe by touch. If this person suddenly gained the ability to see, could they distinguish these objects based on sight alone? This question, now known as Molyneux's problem, has intrigued philosophers for centuries. Between 2007 and 2010, researchers put the problem to the test by studying five children who had regained their vision. The results showed that the children could not match the objects they felt with the ones they had seen.

Gastrointestinal

The stomach is located higher in the body than many people realise. It sits just below the pectoral muscles, less than one centimetre away from the base of the heart.

★ ★

When you blush, your stomach lining also turns red.

★ ★

The world's top competitive eaters are all skinny. This is because thinner people have more room in their abdomen for their stomach to expand, enabling them to consume larger amounts of food. On the other hand, obese people have their stomach constrained by a ring of fat, limiting their ability to eat as much.

★ ★

When you hold in flatulence, some of it can be reabsorbed into your body and released out your mouth.

★ ★

Up to 30% of your body waste is secreted through the sweat glands on your skin.

★ ★

The human enteric nervous system is often referred to as the 'second brain' due to its remarkable complexity. It consists of 500 million neurons—more than the number found in a cat's brain—and operates with its own neurotransmitters, reflexes, and senses. It can also function independently of the brain and spinal cord.

Respiratory

People typically think of the lungs as two hollow, balloon-like pockets of air. However, in reality, they more closely resemble two giant sponges.

★ ★

The lungs contain around 480 million tiny sacs called alveoli, which are located at the end of the respiratory tree. Each one acts like a microscopic balloon, expanding to take in oxygen during inhalation and contracting to expel carbon dioxide during exhalation. These alveoli are the critical sites where gases are exchanged between the lungs and the bloodstream.

★ ★

Your left lung is 10% smaller than your right lung, so as to accommodate your heart.

★ ★

Your body has a clever way of stopping food and liquids from entering your lungs. When you swallow, a small flap called the epiglottis automatically covers your windpipe, directing what you've swallowed down your oesophagus. This action briefly pauses your breathing to protect your airways. For this reason, it's impossible to breathe and swallow at the same time.

★ ★

Every time you take a breath, about 15% of the air in your lungs is replaced with fresh air.

Cardiovascular

Your heart is naturally white. It only appears red because it's filled with blood.

★ ★

Just before you exercise, your heart rate naturally increases in anticipation of the upcoming activity—a process known as the anticipatory rise. A 2022 study found that, due to this phenomenon, a person's heart rate increases by an average of 7.4% more than five minutes before exercise, 12.5% between two and three minutes before exercise, and 24.4% in the final minute leading up to exercise.

★ ★

During a dive, the heart rate of elite freedivers can drop to as low as 11 beats per minute.

★ ★

Urine comes from your blood.

★ ★

Most of the capillaries in your body are so narrow that only one red blood cell can pass through them at a time. In fact, red blood cells often have to deform just to squeeze through.

★ ★

A 2018 study found that patients who had been hospitalised for heart attacks were more likely to survive if the hospital's cardiologist was away at a conference. One possible explanation for this phenomenon is that cardiologists are overusing aggressive treatments, leading to worse outcomes for patients.

The Brain

If all its wrinkles were smoothed out, the brain would have a surface area about the size of a pillowcase.

★ ★

More than half of the brain's surface area is devoted to processing visual information.

★ ★

The brain has no pain receptors, so it cannot feel pain. This is why brain surgery can be performed on patients who are wide awake.

★ ★

Mirror neurons are a type of brain cell that activate when you perform an action and when you see somebody else performing that same action. For example, if you watch somebody pick up a cup and take a sip, the same mirror neurons in your brain will fire as if you were drinking from the cup yourself. Mirror neurons are thought to play a crucial role in helping us empathise with others by mentally 'mirroring' their experiences. In fact, some researchers speculate that dysfunction in the mirror neuron system might contribute to certain conditions, like autism spectrum disorder.

★ ★

The brain is constantly eating itself through a process called phagocytosis. This involves microglia—the brain's primary immune cells—engulfing and digesting cellular debris, pathogens, damaged neurons, and abnormal protein aggregates. This self-maintenance is essential for brain health, much like how pruning a tree helps it grow. However, if it becomes excessive or unbalanced, it can contribute to neurodegenerative conditions, such as Parkinson's, Alzheimer's, or multiple sclerosis.

★ ★

Neurons transmit information throughout your body, enabling essential functions like walking, talking, thinking, and breathing. Unlike most other cells, which undergo a continuous cycle of death and regeneration, neurons are not regularly replaced. They form during early development and can survive for your entire lifetime, making them the longest-living cells in your body. However, this longevity comes with a downside. In neurodegenerative diseases, such as Parkinson's and Alzheimer's, neurons are permanently destroyed and cannot be regenerated. This irreversible loss of neurons makes it extremely difficult to reverse these conditions and restore cognitive function.

★ ★

The brain has a memory capacity of approximately 2.5 petabytes. If it functioned like a computer, that would be enough to store 1.75 billion web pages, ten billion books, or three million hours of TV shows.

★ ★

The brain has the theoretical capacity to store a six-quadrillion-digit-long number.

★ ★

If the brain attempted to store a number with more than 10^{52} digits, it would collapse into a black hole.

★ ★

The brain processes 11 million bits of information every second. However, a person's conscious mind is only aware of about 40 bits of information every second.

★ ★

The brain contains 86 billion neurons connected by 500,000 kilometres of nerve fibre. If this neural network were laid end to end, it would stretch from Earth to the Moon.

★ ★

Information travels through the brain at speeds of up to 400 kilometres per hour. That's faster than a Formula 1 car.

★ ★

Imagination and reality flow in opposite directions through the brain. When you see something in the real world, like a tree, the information flows 'upwards'. It starts in low-level sensory areas (like the visual cortex) and moves to higher brain areas (like the parietal and temporal lobes) to form your perception of reality. However, when you imagine a tree without actually seeing one, this process is reversed, so the information flows 'downwards'. High-level brain areas generate the concept of a tree and send signals to low-level sensory areas (like the visual cortex) to create a mental image.

★ ★

Approximately 4% of the population has aphantasia—a condition in which they cannot visualise images in their mind. To test for aphantasia, close your eyes and try picturing a red apple. Those with strong mental visualisation skills can easily visualise the apple's vivid colour, shape, and even details like the length of its stem. In contrast, those with aphantasia cannot conjure any mental image of the apple at all. Many people with aphantasia are unaware that most of the population can effortlessly create mental images.

★ ★

A 2011 study found a link between brain structure and political orientation. Those with conservative political views had a larger amygdala—a brain region that regulates emotions and assesses threats. In contrast, those with progressive political views had a greater volume of grey matter in the anterior cingulate cortex—a brain region involved in error detection and conflict resolution.

Neurological Phenomena

In 1983, American neuroscientist Benjamin Libet conducted a groundbreaking experiment in which he connected subjects to an electroencephalogram (EEG) and asked them to raise their finger whenever they felt the urge to do so. Surprisingly, the EEG detected brain activity associated with finger movement about 300 milliseconds before the subjects consciously decided to move their finger.

★ ★

In a 2019 study by Roger Koenig-Robert and Joel Pearson, subjects in an fMRI machine were asked to select red or green stripes. By analysing the subjects' brain activity, the researchers were able to predict the colour choice up to 11 seconds before the subjects themselves became aware of their decision. This study, along with Libet's work, raises profound questions about whether humans truly possess free will.

★ ★

If somebody touches your toe and nose at the same time, the signal from your toe will take longer to reach your brain than the signal from your nose. This delay should be noticeable, yet you actually perceive both touches as happening simultaneously. Similarly, if somebody snaps their fingers next to your right ear, the sound will reach your right ear instantly, but it will take a moment to travel around your head to your left ear. Logically, you should hear two distinct snaps, yet in reality, you hear just one. Scientists believe these phenomena happen because the brain waits for the slowest piece of information to arrive, then adjusts the time difference to create the illusion of a single event. This allows us to perceive the world as seamless, even when sensory inputs arrive at different speeds.

★ ★

In a 2002 study, subjects pressed a button that caused a flash of light to appear after a 100-millisecond delay. Over repeated trials, the subjects adapted to this slight delay and began perceiving the flash as happening at the same time as the button press. Then, when the delay was removed—so the flash occurred instantly—something strange happened: the subjects reported that the flash had occurred before they had even pressed the button. This finding demonstrates that the brain continuously recalibrates our sense of time to maintain a coherent perception.

★ ★

People with epilepsy sometimes undergo a surgical treatment called a corpus callosotomy, which severs the main connection between the brain's two hemispheres. Following this operation, each brain hemisphere functions independently, with its own perceptions, impulses, and ideas—almost like there are two brains in one body. Corpus callosotomies have led to some unusual situations. For example, one man who underwent the procedure shook his wife violently with his left hand, while simultaneously using his right hand to come to her aid. In another case, a man who had a corpus callosotomy was an atheist in one hemisphere and religious in the other.

★ ★

Transcranial magnetic stimulation is a non-invasive procedure that uses magnetic fields to stimulate specific brain regions. In some cases, it can temporarily grant a regular person extraordinary skills. For example, it can instantly enable someone to draw well, accurately estimate large quantities, or even solve complex mathematical problems.

★ ★

When transcranial magnetic stimulation is applied to the region of the brain just above the right ear, it can temporarily alter a person's sense of morality.

★ ★

There is some evidence that transcranial magnetic stimulation can enhance a person's sporting ability.

★ ★

Approximately 30% to 40% of people with treatment-resistant depression achieve complete remission by using transcranial magnetic stimulation.

★ ★

When you bend a finger, your brain doesn't send a message just to that one finger. Instead, your pyramidal motor system actually sends a message to all five fingers on that hand, telling them all to bend. At the same time, your extrapyramidal motor system sends a separate message to the other four fingers, instructing them to stay straight. This dual-system coordination is the product of evolution. Our distant ancestors moved all five fingers in unison, but as we evolved, we developed the extrapyramidal motor system, which specifically controls and inhibits the movement of our fingers. The overlap of these two systems has given us the fine motor skills that we have today.

★ ★

Science has virtually no understanding of what consciousness is. We don't know which lifeforms are conscious and which are not. Moreover, as organisms become simpler, we don't know whether consciousness gradually fades or if there's a specific point where it suddenly switches off.

★ ★

Orchestrated objective reduction is a controversial theory proposing that consciousness arises from quantum processes occurring within the tiny structures inside neurons known as microtubules. This idea has been met with scepticism because quantum effects would be expected to quickly break down in the warm, wet, and noisy environment of the brain. However, recent discoveries in quantum biology—such as quantum coherence playing a role in processes like photosynthesis and bird navigation—suggest that it might be possible. Furthermore, tentative research has shown that anaesthetics, which are known to temporarily shut down consciousness, might work by disrupting these delicate quantum states.

Reflexes

Between 18% and 35% of the population experiences the photic sneeze reflex—a condition that causes them to sneeze when they're exposed to bright lights, such as sunlight.

★ ★

The caloric reflex test assesses brainstem function by squirting water into the ear. If cold water is used, the eyes of a person with a healthy brainstem will automatically turn towards the syringe. Conversely, if warm water is used, the eyes will automatically turn away from the syringe.

★ ★

The hanger reflex is a phenomenon in which placing a coat hanger around the head causes the head to involuntarily turn. This occurs because the hanger applies pressure to specific points on the head, triggering the reflex. Interestingly, the hanger reflex is being investigated as a treatment for certain neurological disorders.

★ ★

When cold water touches your face—specifically the area below your eyes and above your cheekbones—it triggers the mammalian diving reflex, causing your heart rate to drop by up to 30 beats per minute. This reflex ensures that oxygen is prioritised for your heart and brain, enabling you to stay submerged underwater for longer.

★ ★

The mammalian diving reflex can be used to reduce anxiety. To activate it, position your face under a cold shower for 30 seconds. Alternatively, hold your breath and submerge your face in a bowl of cold water for 30 seconds. Your parasympathetic nervous system will activate and your heart rate will decrease, leading to a feeling of calmness.

Internal Organs

Your right kidney is lower than your left kidney, so as to accommodate your liver.

★ ★

When a person receives a kidney transplant, their original kidneys are usually left in place, and the new kidney is transplanted into the pelvis.

★ ★

The liver has a remarkable property: it's the only internal organ capable of regeneration. Even if just 25% of its original mass remains, it can regrow into a full-sized, functioning liver.

★ ★

Over the course of a day, the liver can fluctuate in size by as much as 40%.

★ ★

Situs inversus is a rare genetic condition in which the organs in the chest and abdomen are mirrored from their usual positions. In

1979, a Thai woman, Ginggaew Lorsoungnern, survived her first execution because she had this condition. When she was shot at by the firing squad, the bullets missed her heart because it was not in its expected location.

Skin

The skin is surprisingly heavy, making up 15% of a person's total body weight. This means that the average US male has skin weighing about 14 kilograms—equivalent to the entire body weight of an average three-year-old child.

★ ★

Humans are covered from head to toe in stripes called Blaschko's lines. While these stripes are typically invisible, they can become visible on those with certain skin conditions. Blaschko's lines are not random—they are a hidden map of our embryonic development. As we grew from a single cell into a fully-formed human, our skin cells divided, expanded, and stretched, leaving behind these stripes.

★ ★

The human body glows; however, the light it emits is 1,000 times too weak for our eyes to detect.

★ ★

The skin on the elbow has fewer pain receptors than other parts of the body. This is because the elbow's primary function is to provide movement and flexibility, rather than to sense fine touch. As a result, the skin on the elbow effectively doesn't feel pain. You can test this by extending your arm and squeezing the skin dangling from your elbow—you won't feel any pain at all.

★ ★

Your sensory receptors are not evenly distributed across your body. Your fingertips are densely packed with nerve endings and touch receptors, while your forearms contain significantly fewer. For this reason, you can detect fine touches on your fingertips, but not on your forearms. This difference in sensitivity can be illustrated with two experiments. In the first experiment, ask a friend to close their eyes and extend their arm out, palm up. Lightly place your finger on the tip of their index finger and slowly move your finger up their arm. Have them say *"stop"* when they believe you've reached their elbow crease. Most people will say *"stop"* several centimetres too early. In the second experiment, repeat the process, but this time use a paperclip bent in half like a horseshoe, with the two tips spaced one centimetre apart. Ask your friend to say *"stop"* when they can no longer feel two distinct tips and only feel one. Most people will start to feel one tip about halfway up their forearm.

★ ★

When a person's eyes are closed, they typically struggle to identify which of their three middle toes is being prodded. This difficulty arises because the middle toes have a low density of sensory receptors and are positioned so closely together that the touch signals they send to the brain are alike. As a result, the brain receives almost identical inputs from each toe, making it hard to determine which one is being prodded without the help of visual cues.

★ ★

Humans cannot directly sense wetness because our skin lacks the specific nerve endings needed to detect moisture. Instead, our brain makes an educated guess about whether something is wet by interpreting a combination of temperature, pressure, and texture cues.

★ ★

Body wounds sustained during the day heal about 60% faster than those sustained at night.

Hands

The muscles that control finger movement are not located in the fingers—they are actually situated in the palm and forearm.

★ ★

Young children are like starfish: if they cut off the end of their fingertip, it can grow back.

★ ★

Have you ever wondered why your fingertips wrinkle after being in the water? A 2013 study found that wrinkled skin actually improves our grip. This adaptation helped our ancestors by making it easier to grasp plants and animals in wet conditions. However, our fingertips don't remain permanently wrinkled because wrinkling comes at a cost: we lose the sensitivity in our hands.

★ ★

The wrinkling of your fingertips in water is not caused by water soaking into your skin—it's actually a response of your nervous system. When you're exposed to water for an extended period, your autonomic nervous system constricts the blood vessels in your fingertips, causing your overlying skin to fold and wrinkle. For this reason, people with nerve damage in their hands don't experience wrinkled fingertips—their nervous system cannot trigger the response.

★ ★

The fingernails on your dominant hand grow faster than those on your non-dominant hand.

★ ★

Your hands are twice as heavy as you think they are. On average, people estimate that each of their hands weighs about 200 grams. But in reality, the typical human hand actually weighs 400 grams.

★ ★

Most people have three distinct creases on the palm of each hand. However, one in 30 people has just a single crease—a trait that can sometimes be indicative of a developmental disorder.

Bones

The bones in your body are wet.

★ ★

Your bones are not the stark white colour you see in museums and medical illustrations. Living bones actually have a pinkish hue because they contain an extensive network of blood vessels.

★ ★

Textbooks typically claim that the human body contains 206 bones. However, in reality, you almost certainly don't have 206 bones for two key reasons. First, most people have accessory bones—extra bones that form when certain bones fail to fuse properly during development in the womb. Second, as people age, some of their bones naturally fuse, reducing the total number of bones in the body over time. For example, at birth, babies have about 300 bones; by their mid-twenties, most people have about 215 bones; by their mid-forties, this decreases to roughly 209 bones; and by the time people reach old age, this decreases further to about 190 bones.

★ ★

The fabella is a small accessory bone located behind the knee. In 1918, only 11% of the population had a fabella, but by 2018, this figure had risen significantly to 39%. Scientists believe this increase is due to extra pressure on the knee, from people becoming taller and heavier, stimulating the bone's growth.

★ ★

The number of bones and joints in the pinkie toe varies from person to person. About 55% of the population have three bones and two joints, while 44% have two bones and one joint.

★ ★

The hyoid bone, located at the root of the tongue, is the only bone in the human body not connected to another bone. To feel your hyoid bone, place your fingers where your chin and neck meet, then swallow—you'll feel it move.

Muscles

In your body, there is no abrupt boundary where muscle directly meets bone. Instead, the transition between the two occurs gradually in a specialised region called the enthesis.

★ ★

When your muscles contract, they produce low-frequency sounds that are typically inaudible to the human ear. However, under certain conditions, you can actually hear these muscle contractions. To experience this, gently stick your thumbs into your ears, then make a fist. You'll hear a low rumble. The more tightly you clench your fist, the louder the rumble becomes. What you're hearing is the sound of the muscles in your forearms contracting.

★ ★

To hear the sound of your masseter muscle (located in your jaw), rest your head on your hand, so that your palm covers your ear, then start chewing. You should hear a low rumbling—that's the sound of your masseter contracting.

★ ★

In 2021, scientists discovered a new muscle in the jaw.

★ ★

An estimated 40% to 60% of the population has a 24-centimetre-long muscle called the psoas minor. When present, it's found on both sides of the spine, extending from the lower back to the front of the hips.

★ ★

Under normal circumstances, your muscles function at about 60% of their maximum capacity. Your brain imposes this limit to prevent you from tearing your muscles, ligaments, and tendons. However, in extreme life-or-death scenarios, regular people can experience a sudden surge of adrenaline, unlocking what's known as 'hysterical strength'. This extraordinary burst of power enables people to exceed their typical 60% muscular limit. For example, if a child is trapped under a car, hysterical strength might allow the mother to lift part of the vehicle.

★ ★

A 1961 study found that your grip strength increases significantly under certain conditions. It rises by 10% when you're startled by the sound of a pistol, by 15% when you shout or grunt, and by 30% through hypnosis.

★ ★

Grip strength is one of the best indicators of how long a person will live.

★ ★

There is no evidence that stretching before exercise reduces the risk of injury.

★ ★

Sports commentators often describe athletes 'hitting the wall', attributing it to a build-up of lactic acid in their muscles. However, this is a misconception. The human body doesn't actually produce lactic acid—it produces lactate, a similar compound with one less hydrogen ion. Contrary to popular belief, lactate does not cause fatigue. In fact, an experiment demonstrated this by immersing rats' muscles in a lactate concentration equivalent to what would be produced at the end of a 400-metre sprint. The lactate actually had a protective effect on the rats' muscles, rather than a fatiguing effect.

★ ★

Sodium bicarbonate, also known as baking soda, is typically used for cooking and cleaning. However, according to some estimates, 80% of all endurance athletes at the 2024 Paris Olympics took it to enhance their performance. Various studies have found that consuming sodium bicarbonate improves athletic performance by 2% to 3%. It works by neutralising the hydrogen ions that build up during exercise, thereby delaying fatigue.

★ ★

The number of abs you can obtain is determined purely by how many abdominal bands of horizontal connective tissue you're born with. Through exercise, 60% of the population will obtain a six-pack, 20% an eight-pack, 15% a four-pack, and 2% a two-pack.

★ ★

If you were shrunk down to half your size, you'd still be able to jump as high as you can now. This is because your muscle strength (which depends on cross-sectional area) would scale down by a factor of 1/4, while your weight (which depends on volume) would scale down by a factor of 1/8.

Fat

Many professional athletes are classified as obese because the body mass index system cannot distinguish between fat and muscle.

★ ★

Have you ever wondered where your fat goes when you lose weight? Surprisingly, about 84% of the fat you shed is converted into carbon dioxide and exhaled through your mouth. The remaining 16% turns into water and is eliminated as sweat and urine.

★ ★

Over the past four decades, obesity rates have surged so dramatically that more adults on Earth are now classified as obese than underweight. This rise in obesity is partly due to changes in dietary habits and declining levels of physical activity. However, even after accounting for these factors, a significant portion of the increase remains unexplained. For example, a study found that if a person from 2006 followed the same diet and exercise routine as somebody from 1988, the 2006 person would still be about 10% heavier. It's theorised that people have become heavier in recent years due to an increased use of prescription drugs, changes to the human microbiome, and a greater exposure to chemicals.

★ ★

The recent global rise in obesity levels has been linked to an exposure to low doses of pervasive chemicals known as obesogens. These chemicals are commonly found in everyday items, including plastic packaging, food additives, clothing, furniture, and cosmetics. It's believed that obesogens contribute to the development of obesity by directly promoting the growth of fat cells and tissues.

Reproduction

Over the past 50 years, the average male sperm count has halved. In 1973, the typical male had 104 million sperm per millilitre of semen, but by 2018, that number had plummeted to just 49 million. To put this figure into perspective, a sperm count below 40 million significantly reduces the chance of conception. Alarmingly, the decline in the global sperm count appears to be accelerating. If this trend continues, worst-case scenario projections suggest that by 2045, the global sperm count could potentially reach zero. Such an outcome would pose a grave threat to the future of humanity.

★ ★

To facilitate sperm production, the testes must maintain a constant temperature of 34 degrees Celsius. This is why they are located outside the body, away from the average human body temperature of 37 degrees Celsius.

★ ★

The scrotum, which holds the testes, helps regulate their temperature to ensure proper sperm production. In cold conditions, it shrivels and draws closer to the body to keep the testes warm. In hot conditions, it relaxes and hangs lower, away from the body, to keep the testes cool. The scrotum is made of wrinkled skin, enabling it to easily make these expansions and contractions.

★ ★

Over the course of just two weeks, one healthy male produces enough sperm to impregnate every fertile woman on Earth.

★ ★

A baby's gender is determined entirely by the father's sperm that fertilises the mother's egg. If this sperm carries a Y-chromosome, the baby will be male; however, if this sperm carries an X-chromosome, the baby will be female.

★ ★

If your parents had copulated just a few seconds earlier or later, you probably wouldn't exist. Even the slightest variation in ejaculation timing reshuffles the order of the sperm, resulting in a different sperm reaching the egg first.

★ ★

Researchers are developing a fast-acting birth control drug for men, called a sAC inhibitor. It works by temporarily immobilising sperm, preventing them from swimming and fertilising eggs. So far, the drug has only been tested on mice, where a single dose made them infertile within just 30 minutes. It wears off after a few hours.

★ ★

Castration extends a male's life expectancy by an average of 14 years. If the castration is performed before puberty, there is an even greater increase in lifespan. It's believed that castration increases life expectancy by reducing levels of testosterone and DHEAS, both of which are associated with a range of health conditions.

★ ★

Castrated men don't go bald. This is because balding is linked to the sex hormone testosterone, which castrated men produce in just trace amounts.

★ ★

The largest cell in the human body is the egg, while the smallest cell is the sperm.

★ ★

Textbook diagrams typically depict the female reproductive system in a simplified 'exploded' view. The fallopian tubes and ovaries are usually shown stretched out to the sides of the uterus, making it easier to label and understand each part. However, in reality, these organs are all bundled up closely together.

★ ★

The number of eggs required to produce the world's entire human population could fit inside a shoebox.

★ ★

Your maternal grandmother may have carried a part of you in her womb. Females develop the vast majority of their egg cells during the embryonic stage, while still inside their mother's womb. This means the egg that formed you was most likely once inside your grandmother.

Pregnancy

In the earliest known pregnancy test, ancient Egyptian women urinated on barley or wheat seeds. If the seeds sprouted quickly, it indicated that the woman was pregnant. Modern studies have found that this ancient pregnancy test was 70% to 85% accurate.

★ ★

Before modern pregnancy tests became widely available, women commonly used the Hogben test. In this test, a woman's urine sample was injected into an African clawed frog. If the frog laid eggs by the next morning, it indicated that the woman was pregnant. The test was nearly 100% accurate because the pregnancy hormone human chorionic gonadotropin triggered the frog's ovulation.

★ ★

During early development, the gonads are not specific to either gender. By default, they're programmed to develop into ovaries. However, the activation of a single gene, SRY, will trigger a chain reaction of genetic switches that redirects their development into testes. Once the testes form, they produce hormones, such as testosterone, that guide the development of male physical traits. This pivotal process ultimately determines whether the body develops into a male or a female.

★ ★

Have you ever wondered why males have nipples? In the womb, all human embryos—regardless of gender—follow the same developmental pathways until about seven weeks, when the SRY gene kicks in. Because nipple formation occurs so early in development—before the SRY gene is activated—both males and females end up with nipples.

★ ★

You once had a tail. During the fifth week of development, human embryos grow a tail made up of about ten vertebrae. By the eighth week of development, this tail is usually reabsorbed into the body.

★ ★

The vertical groove running from the base of your nose to your upper lip is called the philtrum. It formed when you were an early human embryo. At that time, your face looked quite different: your eyes were situated on the sides of your head; your nostrils and the middle part of your lip were located on the top of your head; and your top lip, jaw, and palate were positioned on your neck. As your face took shape, these three sections needed to grow and merge at precisely the right time. If the timing was off by as little as one hour, you could have developed a cleft lip or palate. This intricate process left no scars, but it did create the distinct groove between your nose and upper lip, known as the philtrum. It's a lasting reminder of how your face came together.

★ ★

The oxygen levels experienced by a fetus in the womb are equivalent to those found on the summit of Mount Everest.

★ ★

A fetus can hear while in the womb. At around 18 weeks, a fetus starts hearing sounds coming from their mother's body, such as her breathing, heartbeat, and digestion. By 28 weeks, their hearing has advanced so much that they begin to hear sounds from outside the womb, like voices and music. At birth, a baby can recognise what they had been exposed to during their final ten weeks in the womb. This includes their mother's voice, her native language, specific types of music, and even nursery rhymes.

★ ★

Babies do not encounter bacteria while in the womb. They're first exposed to bacteria during birth as they pass through the birth canal, which helps establish their early microbiome. Babies born via caesarean section miss out on this crucial exposure to birth canal bacteria, increasing their risk of developing allergies, obesity, and other health conditions. Some researchers are testing whether transferring vaginal fluids to babies born via caesarean section can help replicate the microbial exposure they would have received during a vaginal birth.

★ ★

The placenta is classified as a temporary body organ. It develops in the womb during pregnancy and provides oxygen and nutrients to the baby via the umbilical cord.

★ ★

It's possible to be heavily pregnant and not realise it. In fact, approximately one in every 2,500 pregnancies is called a 'cryptic pregnancy' because it goes unnoticed until labour begins.

★ ★

A woman can give birth when she's dead. This phenomenon, known as coffin birth, occurs when gases build up inside a decomposing body, creating enough pressure to expel the fetus.

★ ★

During pregnancy, most women experience an increase in foot length by two to ten millimetres. In many cases, this change in foot size is permanent.

★ ★

The natural sex ratio at birth is skewed, with approximately 106 male births for every 100 female births. However, in adulthood, the global population evens out to roughly 101 males for every 100 females. This gender ratio balance occurs because males have a higher mortality rate during their youth, due to biological vulnerabilities, health issues, and riskier behaviours.

★ ★

Following earthquakes and other natural disasters, the sex ratio at birth shifts by several percentage points, so that more girls and fewer boys are born. However, during and immediately after wars, the trend reverses, so that more boys and fewer girls are born.

★ ★

The closer a mother lives to the equator, the higher the likelihood that she gives birth to a girl. Conversely, the farther she lives from the equator, the higher the likelihood that she gives birth to a boy.

★ ★

When a man's partner becomes pregnant, his body undergoes its own hormonal and neurological changes to prepare him for fatherhood. For example, his testosterone levels decrease, reducing aggression and promoting caregiving behaviours. Meanwhile, some of his other hormones, like prolactin and oxytocin, increase, strengthening emotional bonds and nurturing instincts. In addition, there is altered activity in several key brain regions associated with empathy—including the prefrontal cortex and amygdala—heightening his sensitivity to the baby's needs.

Babies

The palmar grasp reflex gives newborn babies an incredibly strong grip—so strong that they can support their own body weight while hanging from a bar. The reflex typically fades by around four months of age and is gradually replaced by voluntary grasping—the ability to intentionally reach out and grab objects. Scientists believe the palmar grasp reflex is an evolutionary remnant that once helped infants cling to their mother's fur as she moved and foraged for food.

★ ★

When you firmly stroke the sole of a baby's foot, their big toe bends backwards, while their other toes flare out. This response, known as the Babinski reflex, may be an evolutionary remnant from when our primate ancestors lived in trees and gripped branches with their feet.

★ ★

If you examine the X-ray of a baby's legs, you'll notice something surprising: there are no visible kneecaps. This is because a baby's kneecaps are made entirely of cartilage, rather than bone. Cartilage kneecaps serve two important purposes: enabling a smoother exit through the birth canal and making crawling more comfortable. Between the ages of two and six, the cartilage in the kneecaps gradually hardens and turns into bone.

★ ★

Newborn babies cannot shed tears for the first two to four weeks of their life because their tear ducts are still developing.

★ ★

Newborn babies have extremely limited vision. During their first few weeks of life, they see the world in just black, white, and shades of grey. During their first few months of life, they can only focus on objects 20 to 30 centimetres away—a range that's evolutionary useful for bonding and feeding.

★ ★

Babies cannot look across the midline—an imaginary vertical line that divides the body into left and right halves—until they reach three months of age. Similarly, they cannot reach across the midline until they are six months old. They typically develop the ability to reach across the midline with ease by around four years of age.

★ ★

Thermal synchrony is a phenomenon in which a mother's chest temperature naturally adjusts to help regulate her baby's body temperature. If the baby is too cold, the mother's chest heats up to provide warmth, and if the baby is too hot, her chest cools down to help lower the baby's temperature.

★ ★

Babies cannot taste salt until they are three months old. However, once they develop the ability to taste salt, they actually prefer salty water over plain water.

★ ★

At six months old, human babies can distinguish between individual monkey faces. However, due to a process known as perceptual narrowing, they lose this ability by the age of nine months because it's not a skill they regularly use. Interestingly, if a human baby were raised by monkeys, they would grow up being able to differentiate between monkey faces, but would struggle to tell human faces apart.

Sleep

When you sleep at night, you go through several sleep cycles, each lasting about 90 minutes. For example, if you sleep for seven-and-a-half hours, you'll complete five of these cycles. At the end of each cycle, you enter a phase called rapid eye movement (REM) sleep. During REM sleep, your eyes move quickly in different directions, your arms and legs become temporarily paralysed, and you have vivid dreams. The duration of REM sleep changes throughout the night: it lasts about ten minutes during the first cycle but increases to around one hour by the final cycle.

★ ★

Sleep paralysis is a condition that occurs when REM sleep spills over into wakefulness. In this state, a person is mentally awake but temporarily unable to move their body. During episodes of sleep paralysis, the vivid hallucinations often associated with dreams can occur. This is why some people report seeing ghosts, aliens, or other supernatural phenomena while lying in bed.

★ ★

People sleep worse during a full Moon. This effect holds true even when the person sleeps in a dark room, completely unaware of the Moon's light.

★ ★

A microsleep is a sudden episode of sleep or drowsiness lasting only a few seconds. During this brief period, a person becomes temporarily unconscious and unresponsive to external stimuli.

Alarmingly, most people who experience microsleeps are unaware they have occurred.

★ ★

In a 2014 study, researchers assigned subjects a monotonous task to perform for 50 minutes. During this time, 70% of the subjects experienced at least 36 microsleep episodes.

★ ★

In a 2013 study, researchers used an fMRI machine to monitor the brain activity of sleeping subjects. By analysing this data, the researchers developed an algorithm that could predict the contents of a person's dream with 60% accuracy.

Cancer

A teratoma is a rare type of tumour that can grow its own hair, teeth, muscles, bones, eyes, and even organs.

★ ★

Cancer cells are present in the prostates of 30% of men over 50, 70% of men over 70, and 80% of men over 80.

★ ★

More than two-thirds of Australians are diagnosed with skin cancer at some point in their life.

★ ★

A 2025 study found that getting a tattoo larger than the palm of your hand more than doubles your risk of developing skin cancer and nearly triples your risk of contracting lymphoma.

★ ★

Photodynamic therapy is a cancer treatment that uses a light-sensitive drug and a bright light to destroy abnormal cells. One of the side effects of this therapy is that it can provide people with night vision.

★ ★

The abscopal effect is a rare phenomenon in which a localised treatment applied to a specific tumour leads to the eradication of untreated cancers on other parts of the body. It has the potential to revolutionise future cancer treatments because it demonstrates that properly activating the immune system can destroy cancer cells.

Unusual Conditions

Superior canal dehiscence syndrome is a rare condition caused by a tiny opening in the bone covering the inner ear. This opening disrupts the natural pathways for sound and pressure, causing them to travel abnormally. As a result, people with this condition often experience amplified bodily sounds, such as hearing their own heartbeat or their own eyeballs moving.

★ ★

Reverse-slope hearing loss is a condition in which a person cannot hear low-frequency sounds, such as bass music, the boom of thunder, or the rumble of a car's engine. There is even a documented case of a Chinese woman waking up with this condition and not being able to hear any male voices.

★ ★

Laron syndrome is a genetic disorder in which the body cannot properly utilise growth hormone. As a result, people with this condition don't grow taller than 1.5 metres in height. Remarkably, despite their short stature, those with Laron syndrome are essentially immune to cancer and type 2 diabetes.

★ ★

Stone man syndrome is an extremely rare genetic disorder in which a person's muscles, ligaments, and tendons gradually transform into bone. It's the only known medical condition where one organ system changes into another. Over time, individuals affected by this syndrome effectively turn into statues.

★ ★

Cotard's syndrome is a mental disorder characterised by a variety of extreme delusions. Individuals with this condition often believe they are dead, have lost body parts, or that the world around them no longer exists.

★ ★

Voluntarily generated piloerection is a condition in which a person can give themselves goosebumps on command. It's estimated that one in every 1,500 people possesses this ability. The condition remains scientifically unexplained because the sympathetic nervous system, which controls goosebumps, should function automatically.

★ ★

People with sulfhemoglobinemia—a rare condition characterised by an excess of sulfhemoglobin in the blood—can have blue, green, or even black blood.

★ ★

Tatiana and Krista Hogan are a rare case of conjoined twins fused at the head—a condition known as craniopagus twins. Their brains are connected by a thin band of neural tissue called a thalamic bridge, giving them some extraordinary abilities. They can feel each other's sensations, taste what the other is eating, control each other's limbs, see through each other's eyes, and even hear each other's thoughts.

★ ★

Voodoo death is a phenomenon where a person suddenly dies after experiencing an intense emotional shock, such as extreme fear, grief, or stress. It occurs when the body's fight-or-flight response goes into overdrive, triggering a surge of stress hormones—like adrenaline and cortisol—which overwhelm the heart and other vital organs.

★ ★

Some deaf people with schizophrenia don't hear voices in their head. Instead, they experience visual hallucinations, such as seeing disembodied hands signing to them.

★ ★

There are documented cases of deaf people with Tourette syndrome using sign language to swear involuntarily.

★ ★

Tourette syndrome is often associated with involuntary swearing. However, only about 10% of people with Tourette's actually experience this symptom.

Human Evolution

When you focus intently on a sound, your ears try to turn towards the source of that sound. This happens because your brain sends signals to the tiny muscles around your ears, prompting them to orientate towards the noise. This reflex is a leftover trait from our distant ancestors, who could swivel their ears to precisely pinpoint sounds. Although we lost the ability to freely move our ears around 25 million years ago, these muscles still activate automatically when we concentrate on a sound, sometimes producing a subtle twitch of the ears.

★ ★

The plica semilunaris—the fleshy, pink fold found at the inner corner of the eye—is a remnant of our ancestors' third eyelid.

★ ★

One of the leading theories about why we hiccup suggests that it's a leftover trait from our amphibian ancestors. Hundreds of millions of years ago, our ancestors had both gills for breathing underwater and lungs for breathing air. To prevent water from entering the lungs while submerged, a structure in the throat called the glottis would quickly close. This protective reflex may have stuck with us; so now, when our diaphragm suddenly spasms, our glottis closes, causing us to hiccup.

★ ★

The coccyx is a small, triangular bone at the base of the spine that's a remnant from when our ancestors had tails. As humans evolved and began to walk upright, the tail became unnecessary and gradually disappeared, leaving behind the coccyx as an evolutionary footnote. While the coccyx no longer serves its original purpose, it still provides support to the pelvic muscles and ligaments.

★ ★

The palmaris longus is a small tendon in the forearm that's a remnant from when our ancestors climbed trees. Today, it's absent in about 14% of the population. To check whether you have a palmaris longus, position your hand palm-up, touch your pinkie finger to your thumb, and gently tilt your wrist upwards. If you see a thin, raised, rope-like band poking through the middle of your wrist—that's your palmaris longus. People without this tendon may have weaker pinch strength in their ring and pinkie fingers.

★ ★

Contrary to popular belief, we didn't start out with four feet and have two evolve into hands—instead, we originally had four hands, two of which became feet. During the early stages of primate evolution, our four-legged ancestors transitioned from the ground to the trees, where their four feet evolved into four grasping hands. Later, when our ancestors returned to the ground and began walking bipedally, two of these hands evolved into feet.

★ ★

Humans are covered in the same number of hairs as chimpanzees and other primates: approximately five million follicles. However, human hair is less noticeable than the hair of other primates because it's typically shorter, finer, and lighter in colour.

★ ★

We get goosebumps when we're cold or afraid because they once benefitted our ancestors in these situations. In cold conditions, goosebumps would fluff up our ancestors' fur, providing them with extra insulation to retain body heat. Similarly, when our ancestors were confronted with a threat, goosebumps would make their fur stand on end, so they appeared larger and more intimidating.

★ ★

Female Neanderthals had larger biceps than the average modern-day man.

★ ★

The Bajau people of Southeast Asia have been living on houseboats and diving for food for over 1,000 years. This unique nomadic sea lifestyle has led to their spleens growing 50% larger than normal, allowing them to store more oxygen in their blood for extended dives. As a result of this adaptation, some Bajau divers can now reportedly dive as deep as 60 metres and hold their breath for up to 13 minutes.

Miscellaneous Human Body

Look at a photo of yourself as a child. Is that person truly you? About 98% of the atoms in your body are replaced every year. This means the atoms that once made up your younger self are now scattered around the world.

★ ★

A small amount of the water in your body—about 50 billion molecules—arrived on Earth within the last 1,000 years from beyond the Solar System. This interstellar water hitched a ride on comets, asteroids, and cosmic dust that travelled vast distances before reaching our planet. Over time, it mixed in with Earth's oceans, rivers, and lakes, eventually becoming part of the water cycle and ending up in your body.

★ ★

The Hayflick limit refers to the observation that normal human cells can only divide 40 to 60 times before they die. This biological constraint may explain why the maximum human lifespan appears to be capped at approximately 125 years.

★ ★

The cells in your body are in a continual state of renewal. Your epidermis replaces itself every four weeks, your red blood cells every four months, and your skeleton every ten years.

★ ★

The human body is in a constant state of healing and is never fully healed. A striking example of this can be observed in people who suffer from scurvy. Because they lack the necessary vitamin C levels for collagen production, they may experience old wounds reopening and previously-healed broken bones falling apart.

★ ★

Aristotle famously asserted that humans have five senses: touch, taste, sight, smell, and hearing. However, modern science has discovered that we actually possess up to 33 distinct senses, including the ability to perceive temperature, maintain balance, and sense the passage of time.

★ ★

A 2014 review of 53 trials found that fake elective surgery—involving anaesthetic and incisions but no actual procedure—was just as effective as real surgery in 51% of cases.

★ ★

The average human body temperature in the US has been decreasing by 0.03 degrees Celsius every decade since the 1800s. In

1868, it was roughly 37.0 degrees Celsius, but today it has dropped to about 36.6 degrees Celsius. This drop is believed to be due to modern populations having lower metabolic rates and reduced levels of infection.

★ ★

Normal human body temperature can vary depending on several factors, including age, gender, height, weight, and the time of day. For example, the normal morning temperature of an 80-year-old male who is tall and underweight might be 0.6 degrees Celsius lower than the normal afternoon temperature of a 20-year-old female who is short and obese.

★ ★

Practitioners of g-tummo—a meditation technique that combines specific breathing and visualisation exercises—can self-generate enough heat to increase the temperature of their fingers and toes by as much as 8.3 degrees Celsius.

★ ★

When people share an experience—like watching a live performance or having a conversation—their brainwaves, heartbeat, and breathing all unconsciously synchronise.

★ ★

Contrary to popular belief, there is no evidence that depression is caused by low levels of serotonin.

★ ★

The hiccups have been 'scientifically cured'. A 2004 study was conducted on patients who had hiccups lasting anywhere from 20 minutes to eight hours. The patients were instructed to exhale completely, then take a deep breath in, hold it for ten seconds, inhale a little more, hold it for five seconds, inhale a little more again, and hold it for five seconds. All 16 patients who tolerated this treatment had their hiccups cured.

★ ★

People dry out as they age. Newborn babies are made up of 78% water—a higher percentage than bananas. However, by the time people reach old age, they are made up of just 50% water.

★ ★

Over the past 150 years, the average height of people in industrialised countries has increased by ten centimetres. This is because advancements in nutrition, healthcare, and living conditions have allowed people to achieve their full genetic potential for height.

★ ★

According to some evolutionary biologists, in a few centuries, virtually everyone will look Brazilian.

★ ★

Puberty is beginning earlier than it used to. The average age of puberty in girls has been decreasing by three months every decade since the 1970s. A similar trend, though less pronounced, is occurring in boys. It's not known why this shift is happening. Researchers speculate that rising obesity rates may be playing a role, as well as an increased exposure to chemicals and heightened stress levels.

★ ★

Humans possess several natural tools to detect magnetic fields: we have cryptochrome proteins in our eyes, magnetic material in our nose, and our alpha brainwaves are affected by magnetic fields. Despite these traits, it's not known whether our brain can actually sense magnetic fields or whether magnetic fields impact our behaviour.

★ ★

Women's tears contain an odourless chemical signal that, when sniffed, lowers the testosterone levels in men.

★ ★

The ABCC11 gene determines whether a person produces an underarm odour. Nearly all Koreans carry a variant of this gene that gives them odourless armpits, making deodorant unnecessary.

★ ★

Sweat, by itself, is odourless. Body odour only develops when sweat interacts with bacteria on the skin.

★ ★

The inside of the belly button is connected to the bladder by a tube called the urachus. If this tube doesn't seal off properly before birth, urine can leak out of the belly button.

★ ★

The recurrent laryngeal nerve, which connects the voice box to the brain, takes an unusual detour. Instead of following a direct route, it travels from the voice box down the neck, loops beneath the heart's aorta, and then curves back up to the brain. In giraffes, it takes an even longer detour—descending 4.6 metres down the neck before looping back up.

★ ★

Inner speech—the act of silently talking to yourself in your mind—is accompanied by small muscular movements of the voice box.

★ ★

Research suggests that 5% to 10% of people lack an internal monologue, meaning they don't 'hear' thoughts in their head in the form of silent speech. Studies have found that it's more difficult for people with this condition to perform certain verbal memory tasks, like remembering words or instructions. Those lacking an internal monologue tend to think in images, concepts, or other abstract ways.

★ ★

Approximately one in five people hears a faint sound when they see a flash of light. This condition is caused by a form of synaesthesia, where the brain's sensory pathways cross, causing an overlap between visual and auditory experiences.

★ ★

When someone recognises a familiar person, they often greet them with an eyebrow flash—an involuntary up-and-down movement of the eyebrows lasting just one-fifth of a second. This subtle gesture is generally reciprocated, unless it comes from a stranger.

★ ★

When you consume a drink, your thirst is quenched almost instantly, even though it takes at least five minutes for the fluid to be absorbed into your bloodstream, where it's needed. So, how does your thirst vanish before the drink has had time to alter your blood volume or osmolality? The answer lies in specialised receptors in your throat. When you swallow fluids, these receptors are activated and send an advanced signal to your brain, letting it know hydration is on its way.

★ ★

People in the Southern Hemisphere are more likely to have anti-clockwise scalp hair whorls than people in the Northern Hemisphere.

★ ★

How many times would you need to wrap a string around someone's head to equal their height? Surprisingly, the answer is just three times.

★ ★

More than 80% of schizophrenics smoke.

★ ★

Synovial fluid is a non-Newtonian fluid found in the joints of the body. Its primary role is to minimise the friction between joints during movement, ensuring smooth and painless motion.

Remarkably, synovial fluid is one of the slipperiest substances known to exist, boasting an incredibly low coefficient of friction of just 0.002. To put this value into perspective, ice sliding on ice is less slippery, with a coefficient of friction of about 0.03.

PHYSICS

Physics Phenomena

In 2008, undergraduate student Sebastian Bianchini was preparing a cup of tea when he discovered a curious phenomenon. As he poured water from a kettle into a cup filled with tea leaves, some of the tea leaves seemingly defied gravity and climbed upstream into the kettle. Intrigued, Bianchini shared his observation with two professors at the University of Havana. Together, they conducted further experiments and discovered that when water is poured from a higher container into a lower one, particles from the lower container can climb upwards into the higher one. This counterintuitive behaviour became known as upstream contamination. Although the precise mechanism behind upstream contamination remains a mystery, scientists hypothesise that the surface tension of water allows particles to move upwards against both the flow of liquid and the force of gravity.

★ ★

If you stir a cup of coffee and let it settle, at least one speck of liquid will always end up in the exact spot it started. This phenomenon is a consequence of Brouwer's fixed-point theorem, which states that in any continuous transformation, at least one point must remain unchanged.

★ ★

Have you ever taken a shower and noticed that the shower curtain gets pulled inwards towards the stream of running water? This phenomenon is known as the shower-curtain effect. While several theories have been proposed to explain the effect—including pressure differences, Bernoulli's principle (faster airflow creates lower pressure), and the Coanda effect (airflow follows the curved water stream)—no single explanation has been universally accepted as the definitive cause.

★ ★

Hold a single strand of dry spaghetti at both ends and try to snap it into two pieces. No matter how much force you apply, it will likely break into three or more pieces. This phenomenon is known as the spaghetti-breaking problem. It perplexed some of the world's greatest scientists for decades, including the Nobel Prize winner Richard Feynman. In 2005, the problem was finally solved when scientists discovered that the initial break in the spaghetti triggers vibrations that cause the strand to curve at other points, leading to additional breaks. In 2018, scientists discovered a method to break a piece of spaghetti into exactly two pieces: simply twist the strand by at least 270 degrees before snapping it. The twisting motion counteracts the vibrations, allowing the spaghetti to break cleanly into two.

Acoustics

A sound wave is created when molecules vibrate back and forth through a medium. The stronger these vibrations, the louder the sound. However, there is a physical limit to how loud a sound can be in Earth's atmosphere. That limit is 194 decibels. Beyond this threshold, sound waves compress the air so intensely that they transform into shock waves, which are no longer audible. This occurred when Krakatoa erupted in 1883. It produced the loudest 'sound' in recorded history, estimated at 310 decibels at its source. However, the 'sound' was inaudible because it had transformed into a shock wave.

★ ★

The sound of a rocket launch is so loud that it can damage the rocket and harm the astronauts. To counteract this, 1.1 million litres of water is sprayed onto the launch platform. This water serves as a protective buffer, absorbing and deflecting the rocket's acoustic energy. Contrary to what many assume, the massive white clouds surrounding the base of the rocket during liftoff are not rocket exhaust fumes—they are actually clouds of steam created by this vast amount of water being rapidly heated by the rocket's engines.

★ ★

When you snap your fingers, the sound isn't produced by the top of your thumb hitting your middle finger. It's actually made by your middle finger striking your palm at high speed.

★ ★

If you stand on Westminster Bridge in front of Big Ben, you can hear the clock's bongs on an FM radio before the sound reaches your ears directly. This curiosity occurs because the BBC has installed microphones next to the bells of Big Ben, which capture the sound and broadcast it to your radio. Since radio waves travel faster than sound waves, you can hear the bongs on the radio before they reach you in person.

★ ★

Sonoluminescence is a phenomenon in which a sound wave causes a bubble suspended in liquid to violently collapse and emit a burst of light. During this process, the temperature at the bubble's surface can soar to thousands of degrees Celsius. Some scientists speculate that the temperature inside the bubble may even reach millions of degrees Celsius. That would be hot enough to trigger nuclear fusion—the same process that powers the Sun.

★ ★

Acoustic levitation is a phenomenon in which sound waves are used to suspend an object in mid-air. Researchers have employed this technique to levitate various items, including liquid droplets, polystyrene balls, and even small insects. Acoustic levitation has a broad range of potential applications, including precise drug delivery within the human body, the containerless processing of sensitive or reactive materials, and the contact-free assembly of delicate computer chips.

★ ★

With just ten weeks of training, you can learn to use echolocation to observe your surroundings. To develop this skill, snap your fingers or click your tongue to create sound waves. These sound waves will bounce off the environment, producing an echo. At first, this echo will sound unremarkable, but with regular practice, you'll learn to interpret it. Once mastered, echolocation can allow you to identify the location, size, shape, and even material of nearby objects.

Optics

Contrary to popular belief, the speed of light isn't constant—it varies significantly depending on the medium it travels through. In a vacuum, light travels at 299,792,458 metres per second; when passing through water, it slows down to 226,000,000 metres per second; and through diamond, it's even slower, at 124,000,000 metres per second.

★ ★

The statement *"nothing travels faster than the speed of light"* only applies to light moving through a vacuum. When light moves through any other medium, it slows down. In such cases, a particle can actually surpass the speed of light within that medium.

★ ★

When a particle travels faster than the speed of light in water, it emits a visible blue glow known as Cherenkov radiation. This effect is similar to a sonic boom but involving light rather than sound.

★ ★

Cherenkov radiation is responsible for the characteristic blue glow in the water surrounding nuclear reactors.

★ ★

During cancer radiotherapy, patients often see flashes of blue light, even when their eyes are closed. This blue light is actually

Cherenkov radiation. It's caused by radiation passing through the vitreous fluid in the eye faster than light.

★ ★

Danish physicist Lene Hau has dedicated her career to slowing down light. Using lasers, she can cool a cloud of atoms to just a fraction of a degree above absolute zero, creating an exotic state of matter known as a Bose-Einstein condensate. When light travels through this unique substance, it slows down significantly. In 1999, Hau slowed down light to a mere 17 metres per second. Just two years later, she made history by bringing light to a complete stop.

★ ★

The speed of light is not known with absolute certainty. Scientists have only been able to measure the speed of light for a round trip, such as the time it takes to reach a mirror and return. These experiments provide an average speed of light in two directions. However, the speed of light may be anisotropic, meaning it varies depending on the direction of travel. To date, nobody has managed to design an experiment to measure the speed of light in just one direction.

★ ★

Photons do not experience time. Trillions of years can pass between when a photon is emitted and absorbed, yet no time will elapse for the photon.

★ ★

Stand directly in front of a mirror and mark the top of your head and the bottom of your chin. No matter how far away you move from the mirror, your head's reflection will remain the same size.

★ ★

When two polarising filters—like those found in sunglasses or camera lenses—are aligned at 90 degrees to each other, they block all light from passing through. However, when a third polarising filter is inserted between them at a 45-degree angle, something counterintuitive happens: the light that was previously blocked now passes through all three filters.

★ ★

Place your thumb and index finger close together, leaving a tiny gap of about one millimetre. Hold this narrow slit about five centimetres in front of your eyes. Now, close one eye and look through the slit at a moderately bright object, such as a window, lamp, or computer screen. You should see dark lines appearing within your finger gap. These lines are caused by light producing an interference pattern.

★ ★

In a dark room, shine a laser at a single strand of hair. The light will diffract around the hair, creating an interference pattern on the wall.

★ ★

A laser beam can become trapped in water. This can be demonstrated with a simple experiment. Poke a hole in the side of a water bottle, and allow a stream of water to flow out. Next, shine a laser through the bottle, aiming it at the hole. Instead of continuing in a straight line, the laser beam will curve downwards, following the path of the water.

★ ★

Under the right conditions, gravity can actually create light.

The Electromagnetic Spectrum

The visible light region of the electromagnetic spectrum—which includes everything the human eye can see—makes up just 0.0035% of the entire spectrum.

★ ★

Infrared light is invisible to the human eye. Yet, you can detect it by pointing a TV remote at your phone's camera and pressing any button on the remote. The camera will capture the infrared light, making it appear on the screen. When you try this experiment, use your phone's front-facing camera, as it's less likely to have an infrared filter.

★ ★

Triboluminescence is a phenomenon in which light is created when a material is subjected to mechanical stress, such as rubbing, crushing, or scratching. While most of this light is emitted in the ultraviolet spectrum, which we can't see, some materials produce visible light that appears as a blue or violet glow. To witness the eerie glow of triboluminescence for yourself, try one of the following simple experiments in a dark room: quickly peel the sticky tape off a roll, tear the seal off a glued envelope, open a Band-Aid wrapper, or eat a mint in front of a mirror.

★ ★

In 2008, researchers demonstrated the phenomenon of triboluminescence by rapidly unwinding sticky tape in a vacuum at a speed of three centimetres per second. This action stripped electrons off atoms, producing strong X-rays. These X-rays were so powerful that the researchers were able to use them to take images of the bones in their own hands.

Thermodynamics

The first, second, and third laws of thermodynamics were established in the 19th and early 20th centuries. Later, scientists identified an additional law: if two systems are each in thermal equilibrium with a third system, they must also be in thermal equilibrium with each other. This law proved so fundamental that it needed to precede the other laws. Rather than renumbering the existing laws, the new law was named the zeroth law of thermodynamics.

★ ★

Contrary to popular belief, when you add ice to a glass of water, the ice doesn't cool the water—the water actually heats the ice. This happens because heat always moves from the warmer object (the water) to the cooler one (the ice). This heat transfer continues until both objects reach the same temperature, a state known as thermal equilibrium.

★ ★

Hold a wooden spoon in one hand and a metal fork in the other. Even though the metal fork feels significantly cooler, both objects are actually the same temperature. The metal fork only feels cooler because it's drawing heat away from your skin much faster than the wooden spoon is.

★ ★

Insulating a cylinder, such as an electrical wire, can actually cause it to lose heat faster than if it were left entirely uninsulated. This counterintuitive effect is known as the critical thickness of insulation.

★ ★

A hot cup of coffee has a greater mass than an identical cold cup of coffee.

★ ★

The triple point is the unique combination of temperature and pressure at which a substance can coexist as a solid, liquid, and gas. For water, this occurs at 0.01 degrees Celsius and 611.7 pascals. At this precise point, it exists as liquid water, solid ice, and water vapour, meaning that it can melt, freeze, and boil simultaneously.

★ ★

Try this simple experiment: purse your lips into a small 'o' shape and blow onto your hand. You should feel cool air. Now, open your mouth wide and blow onto your hand. This time, you should feel warm air. In both instances, you're using the same warm air from your lungs, so what's causing the difference? When you blow through pursed lips, the air stream narrows and speeds up. This faster-moving air mixes with the room air, causing it to cool. In contrast, when you blow with your mouth wide open, you produce slower-moving air that doesn't mix with the room air, so it remains warm. To test this further, roll a piece of paper into a tube and blow through it with pursed lips. The air you produce won't mix with the room air, so it will remain warm.

★ ★

Attaining a temperature of absolute zero is impossible. At absolute zero, all the particles within an object would need to come to a complete stop. However, this would violate Heisenberg's uncertainty principle, which states that you cannot simultaneously know a particle's exact position and momentum.

Nuclear

During the early days of nuclear research, some scientists feared that detonating an atomic bomb would trigger a chain reaction of splitting atoms that would set the atmosphere ablaze, destroying the Earth. To address these concerns, the Manhattan Project commissioned a report to evaluate the likelihood of such an event. The report ultimately concluded that this catastrophic scenario was *"unlikely"*.

★ ★

The first atomic bomb test took place in New Mexico on 16 July 1945. The explosion released radioactive isotopes into the atmosphere that had never previously existed in nature. These isotopes, along with those from subsequent nuclear tests, have contaminated virtually every piece of steel produced worldwide since 1945. As a result, highly sensitive equipment—such as Geiger counters, particle accelerators, and spacecraft components—must now be made from pre-1945 steel. Today, the primary source of this pre-1945 steel is World War II-era shipwrecks.

★ ★

On 6 August 1945, an atomic bomb named 'Little Boy' was dropped on Hiroshima, Japan. The bomb worked by splitting uranium atoms through a process called nuclear fission. Each uranium atom that split released a tiny amount of energy—enough to make a grain of sand jump one centimetre. However, as the bomb contained a staggering 2×10^{24} of these splitting atoms, it released a catastrophic explosion.

★ ★

Little Boy released less than 2% of the total energy it could have theoretically released had all its fissile material fully fissioned.

★ ★

In 1952, physicist Ted Taylor lit his cigarette by focusing the intense light from an atomic bomb explosion.

★ ★

On 27 August 1957, a nuclear bomb was placed 150 metres underground at the Nevada Test Site and welded in with a

900-kilogram iron manhole cover. When the bomb was detonated, the immense force caused the manhole cover to fly off at an astonishing speed of 200,000 kilometres per hour—that's 160 times the speed of sound and five times faster than Earth's escape velocity. This made the manhole cover the fastest human-made object ever launched. The manhole cover—or some fragments of it—may have even escaped Earth's gravitational pull and ventured into outer space, becoming one of the first human-made objects to reach the cosmos.

★ ★

A full-body CT scan can expose a person to as much radiation as someone standing just 2.4 kilometres away from the atomic bomb blasts in Japan.

★ ★

Smoking two packets of cigarettes exposes a person to the same radiation dose as a chest X-ray.

★ ★

Employees at the US Capitol Building in Washington, DC, are exposed to higher levels of radiation than nuclear power plant workers. This is due to the substantial amount of uranium present within the building's granite walls.

Motion

If you fire one bullet horizontally from a gun and at the same time drop another bullet from an identical height, both bullets will hit the ground simultaneously. This happens because both bullets are subject to the same gravitational force and will therefore fall at the same rate, regardless of horizontal velocity.

★ ★

In a vacuum, all objects under the influence of gravity fall at the same speed, regardless of their mass, size, or shape. This means that a feather falls at the same rate as a hammer. During the Apollo 15 mission, astronaut David Scott famously demonstrated this principle by dropping a feather and a hammer simultaneously on the Moon. Because the Moon lacks an atmosphere, there was no air resistance to slow either object, hence both objects struck the lunar surface at the same time.

★ ★

Gravity has an infinite range. Its strength weakens with distance, yet it never completely disappears. To illustrate this, imagine two balls—each ten centimetres in diameter and weighing 100 grams—placed one light-year apart in an otherwise empty, static Universe. Despite the vast initial distance between the balls, gravity would slowly drag them together, causing them to collide after 9.5 sextillion years.

★ ★

When an apple falls from a tree, the entire Earth is pulled slightly towards the apple.

★ ★

It's a common misconception that the astronauts aboard the International Space Station float due to a lack of gravity. In reality, they experience 89% of the gravity that we feel on Earth. The reason they float is because they are in a constant state of free fall while in orbit.

★ ★

Gravity is significantly weaker than the other three fundamental forces of nature. It's 10^{38} times weaker than the strong force, 10^{29} times weaker than the weak force, and 10^{36} times weaker than the electromagnetic force. To illustrate gravity's weakness, try using a refrigerator magnet to lift a pin. The magnet's electromagnetic force will overpower the gravitational force of the entire Earth pulling down on that pin.

★ ★

When a bullet is fired from a low angle into water, it bounces off the surface rather than penetrating into it.

★ ★

You can calculate the value of the mathematical constant pi using just two boxes and a wall, assuming a frictionless environment, where no energy is lost. To do this, position a wall on the left, a one-kilogram box in the centre, and another one-kilogram box on the right. When you slide the right box into the central box, the central box will collide with the wall, bounce back, and collide with the right box—sending the right box moving indefinitely to the right. In this setup, the total number of collisions is three, which corresponds to the first digit of pi. Repeating this process, while increasing the mass of the right box by a factor of 100 each time, generates additional digits of pi. For example, if the right box has a mass of 1,000,000 kilograms, there will be 3,141 collisions, and if it has a mass of 1,000,000,000,000 kilograms, there will be 3,141,592 collisions.

Time

In 2020, scientists measured the shortest timespan ever recorded: 247 zeptoseconds. This measurement was made using a laser to track how long it took for a photon to travel across a hydrogen molecule. To put this timespan into perspective, a zeptosecond is one trillionth of a billionth of a second, or 0.000000000000000000001 seconds. The only units of time smaller than a zeptosecond are a yoctosecond and Planck time.

★ ★

There are as many zeptoseconds in one second as there are seconds in a span of time 2,500 times the current age of the Universe.

★ ★

Global Positioning System (GPS) satellites experience time dilation effects due to their high speed and distance from Earth. They travel at approximately 14,000 kilometres per hour, so they lag behind clocks on Earth by 7.2 microseconds per day. However, they also operate farther from Earth's gravitational field, so they run ahead of clocks on Earth by 45.8 microseconds per day. Combined, these time dilation effects result in GPS satellites having a net gain of 38.6 microseconds per day compared to ground-based clocks. If

GPS satellites weren't corrected for time dilation, they would cause errors of up to 11.4 kilometres per day in GPS positioning.

★ ★

Everything you see is in the past. When you extend your arm and look at your fingers, you're actually seeing them as they were three nanoseconds ago. For each additional metre that an object is from you, it takes another three nanoseconds for the light to reach your eyes.

★ ★

The majority of the fundamental laws of physics remain the same, regardless of whether time runs forwards or backwards. For example, if you reverse the direction of time—by replacing t with –t in the equations governing physical systems—nearly all the laws of physics still hold true.

Planck Values

The Planck length, measuring 1.6×10^{-35} metres, is thought to be the shortest measurable distance in the Universe. Any attempt to probe shorter distances through high-energy collisions would result in the formation of black holes. Beyond the Planck length, the very concept of distance becomes meaningless.

★ ★

It would take more Planck lengths to span a single grain of sand than grains of sand to span the diameter of the observable Universe.

★ ★

The strings in string theory are modelled on an order of the Planck length.

★ ★

The Planck time, lasting 5.39×10^{-44} seconds, is the shortest meaningful time interval in physics. It represents the time it takes for a photon moving at the speed of light in a vacuum to cross a distance of one Planck length.

★ ★

There are more Planck times in one second than seconds that have passed since the Big Bang.

★ ★

The Planck temperature, at 1.42×10^{32} kelvin, represents the highest possible temperature in the Universe. At this extreme temperature, our current theories of space and time break down due to the effects of quantum gravity. Scientists believe that the entire Universe was at or near the Planck temperature just 10^{-43} seconds after the Big Bang.

★ ★

While the Planck length, Planck time, and Planck temperature are incomprehensively extreme values, the Planck mass is surprisingly tangible—about 22 micrograms, or roughly the mass of a flea's egg.

Subatomic Particles

To see the atoms in a drop of water with your naked eye, you would need to enlarge the drop to a diameter of one kilometre.

★ ★

If a strand of hair were enlarged to the diameter of the Earth, each atom would be the size of a school bus, each nucleus would be the size of a grain of rice, and each proton would be the size of a grain of salt.

★ ★

The nucleus contains more than 99.9994% of an atom's mass, yet it occupies less than one ten-trillionth of an atom's volume.

★ ★

When people visualise an atom, they typically think of the planetary model, where tiny electrons orbit a central nucleus. However, it's more accurate to think of electrons as a cloud. This cloud represents the probability distribution, showing where an electron is most likely to be found within a given region of space.

★ ★

At this very moment, you are not sitting on a chair; you're levitating above it by some microscopically-sized gap. Similarly, you're not actually holding this book—it's floating just above your fingertips. In fact, throughout your entire life, you have never truly touched anything. This is because, on an atomic level, two objects cannot make direct contact. The electrons in the outer shells of atoms carry identical charges, creating a repulsive force that prevents them from ever touching. What you perceive as touch is actually your skin's nerves detecting this repulsive force.

★ ★

Subatomic particles—like electrons, protons, and neutrons—don't have a colour.

★ ★

When two particles collide in a particle accelerator, they can produce a new particle that has a greater mass than the sum of the original two particles. This is made possible because the immense energy of the collision gets transformed into the mass of the new particle. This occurred in 2012, when the Large Hadron Collider smashed two protons together to create the Higgs boson—a particle 130 times heavier than a single proton.

★ ★

If the human body had just 1% more electrons than protons, each person would carry an enormous negative charge. This charge would be so great that if two people stood at arm's length apart, the repulsive force between them would be strong enough to lift a weight equivalent to the mass of the entire Earth.

★ ★

Have you ever wondered what gives your body its mass? You might be surprised to learn that the mass of the particles that make up your body contribute only a small fraction to your body's total mass. In fact, the vast majority of your body's mass—about 99%—

actually comes from the energy produced by the interactions between the quarks and gluons in your body.

★ ★

Most fundamental forces in nature, like gravity and the electromagnetic force, become stronger as objects get closer together. However, due to a property known as asymptotic freedom, quarks behave differently: their interactions actually become weaker as they get closer together.

★ ★

Electrons might live forever. Scientists estimate that electrons have a minimum lifespan of 66,000 yottayears. That's equivalent to five quintillion times the current age of the Universe.

★ ★

You can build a cloud chamber to detect subatomic particles in your own home. All you need is an empty fish tank, a metal plate, dry ice, and alcohol-soaked felt. Start by attaching the metal plate to the interior-base of the tank, then pour a layer of dry ice onto the plate. Next, attach the alcohol-soaked felt to the tank's ceiling—you'll notice that as the alcohol evaporates and cools, it forms tiny droplets. These droplets will cling to any subatomic particles passing through the tank. In a dark room, shine a torch at the tank, and you'll see ghostly trails. The long, straight trails are created by muons; the short, thick trails come from alpha particles; the forked trails indicate particle decays; and the winding trails are produced by electrons and positrons.

★ ★

Antimatter is the most expensive substance on Earth by an enormous margin. Just a single gram costs an astonishing US$5.3 quadrillion—equivalent to 50 times the value of the entire global economy. The astronomical cost of antimatter stems from the immense challenges in producing and storing it. It can only be created in particle accelerators, which demand vast amounts of energy and resources. Yet, even with these efforts, only trace amounts—often nanograms or less—are made. Antimatter cannot come into contact with normal matter, or it will annihilate, so it must be stored using complex and costly technologies, such as electromagnetic traps in ultra-high vacuum conditions.

★ ★

If a single teaspoon of antimatter came into contact with normal matter, it would produce an explosion large enough to destroy Manhattan.

★ ★

The Milky Way produces nine trillion kilograms of antimatter every second.

★ ★

The laws of the Universe do not apply equally to matter and antimatter.

★ ★

The free will theorem, devised by John Conway and Simon Kochen, states that if humans have free will, then so do elementary particles.

Quantum Mechanics

In 1906, J.J. Thomson was awarded the Nobel Prize in Physics for demonstrating that electrons are particles. Three decades later, his son, George Thomson, received the same honour for proving that electrons are waves. Both men were correct: electrons exhibit particle-like and wave-like properties, in a phenomenon known as wave-particle duality.

★ ★

One of the most famous experiments in physics is the double-slit experiment. The first version of this experiment was conducted by Thomas Young in 1801. Young cut two parallel slits in a metal sheet and directed light through them onto a screen. If light behaved like a particle, it would produce two bright bands on the screen, correlating with the slits. However, if it behaved like a wave, it would pass through the slits, spread out, interact, and produce an interference pattern—featuring a central bright band flanked by gradually dimmer bands on either side. When Young performed the experiment, he observed an interference pattern, meaning that light had wave-like properties. In later variations of the experiment, scientists sent individual photons through the slits, one at a time. Surprisingly, even when individual photons were sent, the interference pattern persisted. In another variant of the double-slit experiment, detectors were used to monitor which slit each photon passed through. When the detectors were added, the interference pattern disappeared, and the photons acted like particles, producing two distinct bands on the screen. The double-slit experiment proves that light exhibits both particle-like and wave-like characteristics, depending on how it's observed.

★ ★

The double-slit experiment isn't just limited to photons. It also works with electrons and even molecules composed of up to 2,000 atoms.

★ ★

A common misconception about the double-slit experiment is that merely observing a photon—by adding detectors—somehow magically changes its behaviour. In reality, when detectors are used to measure which slit a photon passes through, they physically interact with the photon, altering its state. It is this interaction—not the act of observing alone—that causes the photon to behave like a particle, instead of exhibiting a wave-like interference pattern.

★ ★

In 2010, scientists observed the first-ever quantum superposition of an object visible to the naked eye—a 60-micrometre-long metal strip. By cooling and isolating the strip, they made it vibrate and not vibrate at the same time.

★ ★

In 1993, physicists Lev Vaidman and Avshalom Elitzur proposed the quantum bomb thought experiment. In this experiment, a bomb is so sensitive that even a single photon striking it would cause it to explode. The challenge is to determine whether the bomb is live or a dud, without detonating it. In the classical world, testing the bomb without detonation is impossible. However, thanks to quantum mechanics, there is a way. This method relies on the principle of superposition: when a photon encounters two possible paths, it takes both simultaneously. This quantum property enables the photon to detect the bomb's presence without directly interacting with it. While this may sound purely theoretical, real-world experiments—using photon detectors rather than bombs—have successfully demonstrated that superposition allows for interaction-free measurement.

★ ★

Quantum entanglement is a phenomenon in which subatomic particles become linked so that the state of one particle instantly influences the state of the other. Remarkably, this effect holds true, regardless of the distance between the particles. This means that if two particles are entangled, changing a property of one—such as its spin or polarisation—will instantly alter the other, even if it's on the other side of the Universe.

★ ★

Quantum entanglement may sound like science fiction, yet it's a real phenomenon used in various technologies today. In quantum computing, vast numbers of entangled particles work together to solve complex problems that are far beyond the reach of classical systems. In cryptography, pairs of entangled particles are used to create ultra-secure communication channels. If someone tries to intercept one of these channels, the entanglement breaks, immediately alerting both the sender and receiver to the security breach.

★ ★

In 2017, Chinese scientists achieved a record-breaking quantum teleportation distance—they teleported the quantum state of a photon from Earth to a satellite in orbit, 1,400 kilometres away.

★ ★

Quantum tunnelling is a phenomenon in which particles can pass through barriers that would be completely impenetrable in classical physics. To visualise this, imagine throwing a tennis ball at a wall—only for it to suddenly appear on the other side, as if by magic. In the quantum world, this can actually happen because a particle doesn't have a fixed position. Instead, it behaves like a spread-out wave of probability, existing on both sides of the barrier simultaneously.

★ ★

Quantum tunnelling isn't just a theoretical curiosity; it plays a crucial role in a wide range of scientific and technological phenomena. For example, it's integral in the catalytic cycles of enzymes, the functionality of computer flash memory, the development of quantum computers, and the operation of scanning tunnelling microscopes.

★ ★

Quantum tunnelling imposes a fundamental limit on the size of electronic components. For example, when transistors and insulating layers are made less than one nanometre thick, electrons can pass through these barriers by tunnelling. This reduces the device's efficiency, making it more prone to errors.

★ ★

If quantum tunnelling didn't exist, stars wouldn't emit light. Stars produce energy through nuclear fusion, a process that requires two protons to come close enough together to form a nuclear bond. However, because protons are positively charged, they naturally repel each other. Quantum tunnelling allows protons to occasionally overcome this repulsion—once in every 10^{28} encounters. That's the same likelihood as winning the US Powerball lottery three times in a row. Yet, because stars contain such a vast number of protons, these low-probability quantum tunnelling events happen frequently enough to sustain the fusion process and keep stars shining.

Miscellaneous Physics

Diamagnetic levitation is a phenomenon in which non-magnetic materials—such as plastics, water, and even living organisms—float when placed in a strong magnetic field. It occurs because all matter is made up of electrons, and each electron generates its own tiny magnetic field. Normally, these fields are randomly orientated, cancelling each other out. But when exposed to a powerful external magnetic field, they realign to oppose it, thereby creating a repulsive force strong enough to lift the object. In 1997, scientists demonstrated this phenomenon by levitating a frog in a magnetic field that was ten times stronger than a standard MRI machine.

★ ★

Close your eyes and try to visualise what electricity looks like. While we often associate electricity with visible phenomena, like sparks, lightning, and glowing light bulbs, electricity itself is actually invisible.

★ ★

When you flip a switch, the electrons in the wire move towards your appliance at the surprisingly slow speed of just 0.2 millimetres per second—equivalent to 12 centimetres per minute. So, why does your appliance turn on almost instantly? To understand why, think of a wire filled with electrons as a hose filled with marbles. When you push one marble into the hose, it nudges all the other marbles forwards, so one marble pops out the other end. Similarly, when you flip the switch, an electron enters the wire and pushes all the other electrons along. This chain reaction causes an electron to exit the other end of the wire almost instantly, powering your device without any noticeable delay.

★ ★

The probability of all the air molecules in your room spontaneously gathering in one corner due to random motion is approximately one in 10^10^27.

★ ★

A compressed spring has a greater mass than a stretched spring.

★ ★

The University of Queensland's pitch drop experiment is the world's longest continuously run laboratory experiment. It was started in 1927 by Professor Thomas Parnell, who heated a sample of bitumen and poured it into a glass funnel to observe its flow over time. Over the course of nearly a century, just nine drops of the bitumen have fallen. The experiment demonstrates that while bitumen appears solid at room temperature, it is actually an extremely slow-moving fluid, with a viscosity 230 billion times greater than water.

★ ★

If a glass and rubber ball are dropped from the same height, the glass ball will bounce higher than the rubber ball. This is because when a glass ball strikes a surface, its rigid structure preserves most of the impact energy, efficiently transferring it into the bounce. In contrast, when a rubber ball strikes a surface, its flexible structure deforms upon impact, dissipating its energy and resulting in a lower bounce.

★ ★

If you drop Silly Putty from a height, it will shatter like glass when it hits the ground. This is because Silly Putty is a viscoelastic polymer—it typically flows like a thick liquid, but when a sudden force is applied, it hardens into a brittle solid.

★ ★

The Banach-Tarski paradox states that a pea can be cut into a finite number of pieces that can then be rearranged to form a solid ball as large as the Sun.

★ ★

Just four ceramic coffee cups—one placed under each wheel—can support the weight of a two-tonne truck.

★ ★

If you blow out a candle, a plume of what appears to be white smoke will rise from the wick. However, this isn't smoke at all—it's actually vaporised paraffin wax. If you hold a match to it, you can relight the candle without touching the wick.

★ ★

In microgravity, a candle's flame forms a spherical shape and burns with a distinctive blue hue.

★ ★

The Ozma problem poses the following question: if humans were to ever communicate with extraterrestrials from a distant planet, how could we explain the concept of left and right to them without a shared reference point? For centuries, the problem remained unanswered because the physics community assumed that the forces of nature were symmetrical. However, in 1956, physicist Chien-Shiung Wu made a groundbreaking discovery. She determined that the weak force, which governs the decay of subatomic particles, is not symmetrical. When a particle decays, it emits electrons that predominantly spin in one direction. This fundamental asymmetry solves the Ozma problem by offering a way to define left and right universally.

TECHNOLOGY

Computing

Computer companies worldwide produce approximately 842 central processing units every second. In comparison, only 4.3 humans are born every second. This means that society is creating nearly 200 times as many new computers as new people.

★ ★

The metal-oxide-semiconductor field-effect transistor is the most manufactured device in history. Between 1960 and 2018, an estimated 13 sextillion of these transistors were produced. To put this staggering number into perspective, it's roughly equivalent to the number of grains of sand on all the world's beaches.

★ ★

Every year, humanity produces more transistors than grains of wheat and rice combined.

★ ★

The transistors in your computer only work because of quantum mechanics.

★ ★

A greeting card that plays 'Happy Birthday' when it's opened has more computing power than what existed in the entire world in 1950.

★ ★

The TI-83 calculator, released in 1996, has more processing power than the computer used to land Apollo 11 on the Moon.

★ ★

In 1956, a five-megabyte hard drive weighed over one tonne and was so massive that it required a forklift to move. Fast forward to 2025, and a one-terabyte SD card has 200,000 times as much storage capacity, yet it's small enough to fit on your fingertip.

★ ★

The silicon used to manufacture silicon chips needs to be 99.9999999999% pure silicon. This means that for every one billion silicon atoms, there cannot be more than one non-silicon atom present. Achieving this ultra-high purity is crucial because even the tiniest impurities can interfere with the chip's performance.

★ ★

Advanced semiconductors are made using an extremely precise manufacturing process, where some features can be as small as one ten-thousandth the width of a human hair. If even a few atoms are out of place, it can disrupt the semiconductor's electrical properties, potentially causing the entire device to fail.

★ ★

The production of semiconductors requires such a high level of precision that even the smallest of contaminants derailed early research efforts. For example, something as innocuous as touching a copper doorknob would transfer enough copper atoms to a scientist's hand to ruin an entire experiment. Other seemingly trivial factors that disrupted early semiconductor research included scientists' bathroom visits, the menstrual cycles of female researchers, and even the phases of the Moon.

The Future of Computing

The physical scaling of transistors is nearing its limits. As a result, the semiconductor industry is exploring alternative methods to sustain ongoing improvements in computing performance. For example, rather than focusing solely on shrinking transistors, researchers are investigating whether they can stack transistors vertically to increase packing density. Other researchers are looking into advanced manufacturing techniques, such as extreme ultraviolet lithography. This involves using high-energy ultraviolet light to print tiny circuit patterns on chips, enabling the finer patterning of transistors. Additionally, there is a growing interest in developing semiconductors from novel materials, like graphene and carbon nanotubes, which could enhance computing performance. Other efforts include designing specialised chips that don't rely on transistor density, such as AI accelerators and quantum computer systems.

★ ★

Classical computers process information using bits, which can represent either 0 or 1. In contrast, quantum computers process information using qubits, which can exist simultaneously in a superposition of the 0 and 1 states. This unique property makes quantum computers exponentially more powerful than classical computers. For example, a quantum computer with just 300 qubits would theoretically be more powerful than all the classical computers in the world connected. In fact, such a quantum computer could perform more calculations in a fraction of a second than there are atoms in the observable Universe.

★ ★

In 2024, Google unveiled a 105-qubit quantum computing chip named Willow. It completed a specific task in five minutes that it would take the world's most advanced supercomputer ten septillion years to solve.

★ ★

Scientists have demonstrated that it's possible to build a quantum computer that operates using sound particles, known as phonons.

★ ★

Neuromorphic computing is an emerging field of computer science focused on developing computers that emulate the biological processes of the human brain. Unlike traditional computers, which process tasks sequentially, neuromorphic systems use artificial neurons and synapses to mimic the brain's parallel and energy-efficient communication architecture. This approach significantly reduces power consumption, while excelling at complex tasks like pattern recognition, sensory processing, and decision making.

★ ★

Researchers at Western Sydney University have developed DeepSouth—the world's first neuromorphic supercomputer. It mimics the brain's neural network by simulating vast arrays of neurons and synapses. DeepSouth can perform 228 trillion synaptic operations per second—a processing speed comparable to that of the human brain.

Artificial Intelligence

The decision-making processes of complex artificial intelligence systems—like large language models and deep neural networks—are not entirely understood by anyone, even their creators.

★ ★

Artificial intelligence can detect abnormalities in electrocardiogram results, even when they look normal to doctors. It's not known how the AI achieves this.

★ ★

Artificial intelligence can accurately estimate a person's IQ by analysing their brain circuitry in action.

★ ★

Ray Kurzweil is a computer scientist and futurist who is widely renowned for his accurate technological predictions. He envisions that by the 2030s, a hybrid artificial intelligence will emerge that blends human and machine capabilities. This system would connect the internet directly to the human brain, allowing us to instantly download skills, knowledge, and even memories.

★ ★

The AI singularity refers to a hypothetical future point at which artificial intelligence becomes so advanced that it can undergo recursive self-improvement. This means that it could enhance its own intelligence autonomously, without human intervention. Once this point is reached, technological growth may accelerate beyond human control, becoming unstoppable and irreversible. This could lead to machines asserting significant influence over humans, resulting in profound consequences for humanity.

★ ★

Ray Kurzweil predicts that the AI singularity will be reached by 2045, resulting in the emergence of non-biological intelligence that is one billion times more powerful than the combined intelligence of all humans today.

★ ★

Artificial intelligence researcher Leopold Aschenbrenner predicts that by 2030, running a single AI training cluster will require more than 20% of the United States' total electricity production. This staggering energy demand could significantly hinder the rapid growth of AI.

Phones

Next time you're on a phone call, listen closely—you should be able to hear a faint layer of artificial background noise, known as comfort noise. It's intentionally added to phone calls to fill in the silence and reassure both callers that their connection hasn't dropped out.

★ ★

In 2013, Motorola developed an edible 'password pill' that turns your body into a living password. The pill contains a tiny chip that activates when it comes into contact with your stomach acid. Once activated, the chip emits an 18-bit signal that unlocks your smartphone and computer.

★ ★

Scientists can estimate your approximate age by analysing the fingerprint smudge marks on your phone's touchscreen.

★ ★

Touchscreen technology only works because of your body chemistry. The sensors in your phone's touchscreen are designed to ignore poor conductors, like gloves, rubber, and fingernails, as well as excellent conductors, like metal. They are programmed to only respond to moderate conductors, such as a human finger. However, these sensors cannot distinguish between a human finger and an object with a similar conductivity, like a banana. For this reason, you can use a banana to operate your phone's touchscreen.

★ ★

Smartphone technology relies on roughly 250,000 different patents.

★ ★

It's possible to charge your mobile phone using the energy generated from a fire, train tracks, or even a hand-crank flashlight.

★ ★

Scientists have developed a material called magnetoelectric multiferroic, which could make a smartphone battery last for three months on a single charge.

Satellites

Satellites do not track GPS devices; GPS devices actually track satellites. Each GPS satellite in orbit continuously broadcasts a signal indicating its own precise location. GPS devices on Earth receive these signals from at least four satellites, measure their distance to each one, and then use this data to calculate their own exact position.

★ ★

One-third of all the tracked space debris currently orbiting Earth originated from just two events: the intentional destruction of a Chinese weather satellite in 2007 and the accidental collision between an American and a Russian satellite in 2009.

★ ★

An estimated 130 million pieces of untracked space debris, ranging in size from one millimetre to one centimetre, are currently orbiting Earth. These fragments are mostly remnants of broken satellites and discarded rocket components. They are too small to be tracked, yet they typically travel at speeds exceeding 28,000 kilometres per hour, making them potentially dangerous. In 2016, one of these pieces—believed to be a speck of paint only a few thousandths of a millimetre wide—caused noticeable damage to a window on the International Space Station.

★ ★

Approximately 10,000 satellites are currently orbiting Earth. However, roughly one-third of them are non-operational and considered space junk.

★ ★

It's estimated that more than 100,000 satellites could be orbiting Earth by 2030.

★ ★

The Kessler Syndrome refers to the potential scenario where collisions between space debris trigger a runaway chain reaction that continuously generates more debris. Over time, this could result in Earth becoming completely surrounded by a dense cloud of space junk. Such an outcome would not only prevent the launch of new satellites, thereby disrupting critical services like communications and weather monitoring, but also effectively trap humanity on Earth by making space travel impossible.

Space and Technology

In the distant future, when all the stars have died and only black holes remain, an advanced civilisation's only means of survival may be to harness energy from a rotating black hole. To achieve this, they would need to construct a mirror that completely surrounds the black hole. By directing electromagnetic waves into this system, the waves would repeatedly bounce between the mirror and the black hole, growing exponentially stronger over time. If the civilisation then opened the mirror, they would gain trillions of years' worth of energy. Applying this technique to the supermassive black hole at the centre of the Milky Way would yield an amount of energy equivalent to the total output of every star in the Milky Way combined across one billion years. Alternatively, if the civilisation chose to keep the mirror closed, they would have created a 'black hole bomb', which, if detonated, would release as much energy as a supernova.

★ ★

Black holes played a role in the development of Wi-Fi. In 1974, physicist Stephen Hawking proposed that primordial black holes—hypothetical black holes smaller than an atom, but with a mass of Mount Everest—might emit radio waves when they evaporate. Inspired by this idea, Australian engineer John O'Sullivan developed a sophisticated mathematical tool to isolate these faint radio waves from the background noise. While this tool failed to detect any signs of primordial black holes, it proved invaluable years later when O'Sullivan turned his attention to a new challenge: enabling wireless communication between computers. Like the elusive black hole radio waves, wireless signals were weak and buried in background noise. O'Sullivan took the mathematical tool he had developed to detect black hole radio waves and adapted it to strengthen wireless communication signals. Remarkably, it worked. This serendipitous breakthrough led to the development of Wi-Fi.

★ ★

NASA emails tools to astronauts aboard the International Space Station, which are then printed out on the space station's 3D printer.

★ ★

Russia once planned to create an artificial Moon. They envisaged that it would direct the Sun's rays onto the dark side of the Earth, enabling people to work at night. In 1993, Russia took a step towards their ambitious goal by launching a 20-metre-diameter mirror into orbit. The mirror successfully directed a beam of light twice as bright as the Moon onto a five-kilometre-wide patch of the Earth's surface. This bright spot swept across Europe at a speed of eight kilometres per second. After several hours, the mirror was de-orbited and burned up in the atmosphere. Russia had plans to launch an even larger, 70-metre-wide mirror into orbit, but the project was ultimately abandoned.

★ ★

California-based company Reflect Orbital wants to sell you sunlight. In 2024, they announced plans to launch a large mirror into space that would reflect the Sun's rays onto specific locations on the Earth's surface. This would provide you with sunlight on demand during the night.

★ ★

In 1993, US-based company Space Marketing Inc. planned to launch a one-kilometre-wide billboard advertisement into space. Their goal was to make the billboard appear as large and as bright as the Moon, orbit Earth for two weeks, and be visible to half the world. However, their proposal was ultimately blocked by the US House of Representatives.

★ ★

Helium-3 is an isotope that is extremely rare on Earth but found in abundance on the Moon's surface. It's highly sought after because it has the potential to fuel nuclear fusion reactors, providing a source of clean energy that produces no radiation and virtually no waste. It's so efficient that some scientists estimate that just three space shuttle missions to the Moon annually could yield enough helium-3 to power the entire planet. In 2024, US-based company Interlune announced plans to mine the Moon's vast helium-3 supply and return quantities to Earth by 2030.

★ ★

Researchers are developing a GPS-like system designed for deep-space navigation called XNAV. It operates by using the X-rays emitted by pulsars to pinpoint a spacecraft's location to an accuracy of five kilometres.

Space Transportation

In 1999, NASA permanently lost contact with the Mars Climate Orbiter because of a rudimentary error: the engineers working on it failed to convert the measurements from English units to metric.

★ ★

In 2015, NASA's New Horizons spacecraft flew past Pluto after a nine-year journey, covering 5.3 billion kilometres. Remarkably,

NASA's prediction of how long the journey would take was accurate to within one minute.

★ ★

In 2023, the Parker Solar Probe became the fastest human-made object in history. Utilising Venus as a gravity slingshot, it reached an astonishing speed of 635,266 kilometres per hour. To put this speed into perspective, it's 210 times faster than a speeding bullet and quick enough to travel from London to New York City in just 32 seconds.

★ ★

Space probe Voyager 1 has been travelling at roughly 17 kilometres per second for almost five decades. It's now located approximately 25 billion kilometres away from Earth, making it the most distant human-made object in space. While this distance may seem immense, it's only one-light-day away.

★ ★

About 85% of a rocket's mass is taken up by the propellant that's used to launch it into space. The remaining 15% constitutes everything else: the rocket's structure, engines, and payload.

★ ★

During launch, the SpaceX Starship generates around 131 gigawatts of power. To put this immense number into perspective, it's equivalent to 10% of the power output of the entire United States grid.

★ ★

Most of a rocket's engines aren't designed to propel it vertically upwards—their main role is to provide the horizontal velocity needed to achieve and maintain orbit. When a rocket is launched, it first travels vertically to escape Earth's dense atmosphere and reduce drag. Once it clears the thickest part of the atmosphere, it performs a manoeuvre called a gravity turn, where it gradually tilts and accelerates horizontally. To remain in orbit, it must reach a horizontal velocity of about 28,000 kilometres per hour. The majority of a rocket's engines and fuel are dedicated to reaching this critical speed.

★ ★

With today's technology, reaching Proxima Centauri, the closest star to the Solar System, would take a spacecraft 6,300 years. To ensure that a healthy and genetically diverse population arrived at the destination after such a long journey, a starting crew of at least 98 people would be necessary.

★ ★

If a spacecraft could maintain a constant acceleration of 1 g, it would be possible to travel across the entire observable Universe within a single human lifetime.

★ ★

In 2024, NASA launched the Advanced Composite Solar Sail System into Earth's orbit. This innovative sail, which is roughly the size of a small apartment, has a remarkable means of propulsion: it uses the momentum of photons. The development of this technology could pave the way for future deep-space missions that don't require rocket fuel after launch.

★ ★

In 2016, physicist Stephen Hawking, alongside billionaires Yuri Milner and Mark Zuckerberg, co-founded Breakthrough Starshot. This ambitious project plans to send a fleet of 1,000 postage-stamp-sized spacecraft to the Alpha Centauri star system, 4.3 light-years away. Each miniature spacecraft would be equipped with a reflective sail measuring four metres across and just a few hundred atoms thick. A powerful ground-based array of lasers would fire at each sail for ten minutes, propelling the spacecraft. If technology continues to progress at its current rate, the spacecraft would travel at approximately 20% of the speed of light. At this speed, the journey to Alpha Centauri would take 22 years. Upon arrival, the spacecrafts' onboard cameras would capture images of the star system, although it would take an additional four years for this data to reach Earth. It's estimated that the first Breakthrough Starshot spacecraft could be launched as early as 2036.

★ ★

US-based company SpinLaunch plans to throw a rocket into space. The company has developed a vacuum-sealed centrifuge that's able to spin a rocket at several times the speed of sound. Upon release, the rocket can reach speeds of up to 7,500 kilometres per hour. Following this launch phase, booster rockets would kick in and take the payload the remaining distance into orbit. If SpinLaunch's technology proves successful, it could slash the cost of reaching space by a factor of 20.

★ ★

A space elevator is a proposed transportation system that offers a practical and cost-effective means to reach space. The main component of this system is a cable extending 100,000 kilometres into space—about one-quarter of the distance to the Moon. One end of the cable would be anchored to Earth's surface, while the other end would be attached to a counterweight to maintain tension. A mechanical climber would travel up and down the cable, transporting passengers and cargo into orbit. The greatest challenge in building a space elevator lies in finding a suitable material to construct the cable. This material would need to be incredibly strong, lightweight, and capable of being produced in large quantities. The top three candidates for such a material are single-crystal graphene, carbon nanotubes, and hexagonal boron nitride. With advancements in material science, the world's first space elevator could be constructed in the next few decades.

★ ★

In 1994, physicist Miguel Alcubierre proposed a theoretical concept for faster-than-light travel called the Alcubierre drive. This hypothetical spacecraft would achieve faster-than-light speeds by compressing the space in front of it and expanding the space behind it, effectively creating a 'warp bubble'. Since the spacecraft itself wouldn't be moving through the local space faster than light, it wouldn't violate Einstein's theory of relativity. Initially, the concept of an Alcubierre drive seemed implausible because generating the necessary negative energy to power it would require more mass than what exists in the entire observable Universe. However, recent research suggests that an Alcubierre drive might be possible without the use of negative energy.

★ ★

We may soon be able to detect whether any extraterrestrial life in our galaxy is using warp drive technology. A 2024 study found that when an Alcubierre drive malfunctions, it emits a distinct burst of gravitational waves. The next generation of gravitational wave detectors on Earth might be sensitive enough to detect these waves.

★ ★

According to a 2023 paper, it might be possible to escape from inside a black hole using an Alcubierre drive.

★ ★

The Wait Calculation suggests that humans should refrain from partaking in interstellar travel unless we can reach the destination within 50 years. The reasoning is that if we embark on an interstellar voyage with today's technology, a future mission, equipped with more advanced technology, would still arrive at the destination first.

Transportation

Scientists don't fully understand the precise mechanism by which planes stay in the air. However, engineers can still successfully design aircraft because they know what key characteristics will keep them aloft.

★ ★

Commercial airliners today take roughly 10% longer to reach their destination than they did in the 1960s. This is primarily because modern aircraft are designed for fuel efficiency rather than speed.

★ ★

In 2024, Mercedes-Benz unveiled a highly efficient solar-powered paint that can generate electricity when applied to vehicles. Just a single coat of this paint applied to a medium-sized SUV can produce enough energy to power the vehicle for up to 20,000 kilometres annually.

★ ★

When an electric car travels under 30 kilometres per hour, the sound you hear isn't generated by the vehicle. Instead, it comes from a pre-recorded sound file played through external speakers. This artificial noise is added to alert nearby pedestrians that a car is in the vicinity.

★ ★

Formula 1 cars generate such immense downforce that they can drive upside down on the ceiling of a tunnel.

★ ★

Sailboats can reach speeds four times faster than the wind that propels them.

★ ★

If you tied a piece of string between the two interior walls of a submarine and then submerged deep underwater, the string would slacken. This is because the submarine's outer shell would compress under the extreme pressure of the deep ocean, causing the distance between the walls to decrease.

★ ★

Scientists don't fully understand why a bicycle remains upright and balanced while in motion.

★ ★

Wearing a helmet as a cyclist is widely regarded as a sensible safety measure, yet the idea of a pedestrian or driver wearing a helmet seems absurd. However, research shows that wearing a helmet in each of these three scenarios provides a similar benefit. For instance, a 1996 Australian study found that cyclists have a head injury risk of 0.41 per million hours travelled, while motor vehicle occupants and pedestrians actually face higher risks of 0.46 and 0.80, respectively. Similarly, a 2006 French study found that pedestrians are 1.4 times more likely to suffer a traumatic brain injury than cyclists who don't wear a helmet.

Construction

Lego bricks can withstand compression better than concrete. In fact, a single Lego brick can support the weight of 375,000 other Lego bricks before it fails.

★ ★

Sand is the world's second-most-consumed natural resource, after water. It's a crucial ingredient in the production of concrete, asphalt, glass, and silicon chips. However, despite its abundance, the world is facing a sand shortage. This issue has arisen because not all sand is the same. Desert sand, which is widely available, is too smooth and rounded to be useful for construction. In contrast, riverbank and shoreline sand, with its angular and coarse grains, is perfect for construction. The high demand for river and coastal sand has led to its over-extraction, leaving many riverbeds and beaches stripped bare.

★ ★

Scientists have developed self-healing concrete—a material that automatically repairs its own cracks.

★ ★

Researchers have created transparent wood, a material that is stronger and lighter than conventional glass, while also offering superior insulation. They achieved this by removing the lignin from wood and replacing it with a polymer. This made the wood transparent while preserving its natural structure. Transparent wood has the potential to revolutionise the construction industry, particularly the development of solar panels and energy-efficient windows.

★ ★

It's projected that by 2030, one-quarter of all new buildings in Dubai will be 3D printed.

★ ★

When engineers design large-scale structures, they must take into account the curvature of the Earth. A notable example of this is New York City's Verrazzano-Narrows Bridge, which has two

211-metre-tall suspension towers spaced 1.3 kilometres apart. Because the Earth is curved, the distance between the tops of the towers is 4.1 centimetres greater than the distance between the bases of the towers.

★ ★

Many skyscrapers use an advanced weight system to minimise swaying during high winds. When strong winds are detected, a computer adjusts massive counterweights—each weighing several hundred tonnes—positioned on the skyscraper's top floor. By strategically shifting these weights, the system counteracts the wind's force and stabilises the building.

★ ★

A 2020 study found that human-made materials, such as concrete, metals, and plastic, now collectively weigh more than all the living biomass on Earth.

Military

The US military has developed microwave weapons that can be used for riot control, perimeter security, and convoy protection. They work by emitting microwave energy that heats the water molecules in a person's skin, causing an intense sensation of pain. This pain forces the person to move away from the area, making the technology an effective, non-lethal tool for dispersing crowds and deterring threats.

★ ★

It's possible to remotely disable every nuclear bomb in the world. To achieve this, you'd need to fire a high-energy neutrino beam of around 1,000 teraelectronvolts at the Earth. This beam would trigger fission reactions in every nuclear bomb on the planet, effectively melting or vaporising them. However, constructing such a beam would cost in excess of US$100 billion and require over 50 gigawatts of power—more than the total power output of Great Britain.

★ ★

In 1961, the Soviet Union tested the Tsar Bomba, the most powerful device ever deployed on Earth. With a yield of 50 megatons, it was ten times more powerful than all the ordnance exploded during World War II combined. The bomb was detonated over the Arctic archipelago of Novaya Zemlya, producing a blinding flash that was visible over 1,000 kilometres away in Alaska, Norway, and Greenland. The explosion created a massive mushroom cloud that rose 67 kilometres into the atmosphere—seven times the height of Mount Everest—and generated a seismic shock wave that circled the globe three times. The thermal radiation from this detonation was so intense that it would have inflicted third-degree burns on anyone 100 kilometres away.

★ ★

The US Navy has developed electromagnetic railgun technology capable of accelerating projectiles to a speed of Mach 7 almost instantaneously. This advanced weapons system eliminates the need for traditional explosives, relying instead on kinetic energy to deliver precise, high-impact strikes over long distances.

★ ★

The US military has created exoskeleton technology—a wearable robotic suit that enhances a soldier's physical capabilities. This technology boosts a soldier's strength, increases their endurance, and allows them to carry heavier loads with less strain.

★ ★

In 2017, scientists equipped a genetically engineered dragonfly with a solar-powered backpack and steering implants. This effectively turned the dragonfly into a living drone.

Surveillance

Persistent Surveillance Systems has developed an innovative technology that can solve crimes from the sky. The system uses a discreet plane that circles above a city for up to ten hours at a time. A powerful camera attached to the plane records the 80-square-kilometre area below. The system works like a real-time version of Google Maps, so when a crime is committed, the authorities can review the footage to identify the perpetrator, determine where they've come from, and where they've travelled to. The technology has already been used in various cities across the US and Mexico, helping the police solve major crimes in a matter of minutes.

★ ★

A laser microphone is a surveillance device that can be used to eavesdrop on conversations. When a person speaks, their voice creates sound waves that cause nearby objects to vibrate by a matter of nanometres. A laser beam can detect these subtle vibrations, and a receiver can convert them back into audible sound. In 2011, the

US military used this technology to monitor Osama bin Laden's compound in Abbottabad, Pakistan. They directed a laser beam at the window of the compound, analysed the vibrations, and determined that 22 people were inside the compound, including one person who never stepped into view.

★ ★

A 2014 experiment found that it's possible to eavesdrop on conversations by videoing the subtle vibrations of nearby inanimate objects. In the experiment, a person recited the nursery rhyme '*Mary Had A Little Lamb*' near a potato-chip packet, causing the packet to vibrate slightly. A camera, positioned five metres away behind soundproof glass, recorded these nanometre-sized vibrations. By analysing the video, the researchers successfully converted the vibrations into a clear audio reconstruction of the nursery rhyme. They achieved similar results by videoing the vibrations of other objects, including plant leaves, aluminium foil, and the surface of a glass of water.

★ ★

Artificial intelligence can accurately identify which keys are being pressed on a keyboard simply by analysing the sound of someone typing. In experiments, AI systems achieved accuracy rates of 92% when analysing sounds over Skype, 93% over Zoom, and 95% over phone calls.

Audio

Gunfire detection technology is a cutting-edge crime-fighting tool that uses sensors equipped with a microphone, GPS, and cellular activity to accurately pinpoint the location of shootings. When a gunshot occurs, the sensors—distributed at a density of 50 per square kilometre—detect the sound, triangulate its precise location, and create a map of the incident, which is immediately relayed to the police. The technology has already been adopted by more than 130 cities and law enforcement agencies in the US.

★ ★

During the 2010s, some television commercials emitted ultrasonic tones that were inaudible to the human ear but could be picked up by nearby smartphones and tablets. When devices detected these sounds, companies could link a user's activity across multiple platforms—such as TVs, phones, and tablets—using browser cookies. This allowed companies to track whether somebody viewed a commercial, how long they watched it, and if they purchased the advertised product.

★ ★

The microwave auditory effect is a phenomenon in which microwave pulses directed at a person cause them to perceive sounds—such as clicks, buzzes, or even words—directly inside their head. This technology works because the microwave pulses heat brain tissue, creating pressure waves that are interpreted as sound. The US military has conducted extensive research into microwave pulses, exploring their potential use as a warfare tool.

★ ★

Scientists are developing an acoustic fire extinguisher that uses sound waves to put out forest fires. Rather than relying on water or chemicals, this device generates low-frequency sound waves that disrupt the flames by separating the oxygen from the fuel—effectively snuffing out the fire.

Microwaves

An unplugged microwave contains enough residual charge to kill you.

★ ★

Have you ever wondered why your microwave's front screen is covered in small black dots? These dots are spaced close enough together—typically much smaller than the microwave wavelength—to contain microwaves within the chamber. At the same time, they are spaced far enough apart to allow visible light through, so you can see your food as it cooks. If this mesh screen becomes damaged, you may be exposed to dangerous radiation levels.

★ ★

Most people don't use their microwave correctly. People tend to place their food in the centre of the rotating plate; however, this is actually the worst place to put it because it remains static in the centre of the microwave, resulting in uneven heating and cold spots within the food. For best results, you should position your food near the edge of the rotating plate. This allows it to rotate more effectively, ensuring that the microwaves are evenly distributed and your food is heated throughout.

★ ★

You can calculate the speed of light using your microwave. Start by removing your microwave's rotating tray and replacing it with a plate covered in cheese. Turn the microwave on, and the waves inside will bounce around until they create a standing wave consisting of nodes (points of low energy and poor heating) and anti-nodes (points of high energy and effective heating). When you remove the plate, you'll see hot and cold spots on the cheese that correspond to the locations of these nodes and anti-nodes. The distance between any two adjacent nodes or anti-nodes equals half the wavelength of the microwave. If you multiply this wavelength by the microwave's frequency—this value is displayed on a sticker on the microwave—you will have calculated the speed of light.

★ ★

A typical microwave uses roughly the same amount of electricity powering its clock as it does heating food.

Nanotechnology

Nanotechnology is a branch of technology focused on manipulating matter at the nanoscale. This scale is one thousand times smaller than the microscopic level and one billion times smaller than a metre. At the nanoscale, quantum mechanical effects come into play, causing materials to exhibit unique physical, mechanical, chemical, and optical properties. These special characteristics make nanotechnology useful in various fields, including medicine, electronics, energy, and environmental science.

★ ★

In 2010, Andre Geim and Konstantin Novoselov were awarded the Nobel Prize in Physics for discovering graphene and identifying its unique properties. Their method of discovery was surprisingly simple: they used adhesive tape to peel away layers of graphite, the same material that makes up pencil lead. Repeating this process, they successfully isolated flakes just one carbon atom thick, creating the world's first graphene. This nanomaterial exhibits some extraordinary properties. For instance, a one-square-metre sheet weighs as much as a strand of hair and is invisible to the naked eye, yet it's strong enough to support the weight of a four-kilogram cat.

★ ★

Nanobots are microscopic robots designed to carry out specific tasks at the nanoscale. While still in the research and development phase, they have the potential to revolutionise medicine. In the near future, they could be used to deliver drugs to specific parts of the body, repair damaged organs, destroy cancer cells, and even slow the aging process. Futurist Ray Kurzweil predicts that by 2030, some people will have nanobots circulating in their bloodstream, actively preventing and treating disease.

★ ★

In 2024, scientists made a breakthrough in nanotechnology—they used nanorobots to shrink bladder tumours in mice by 90%. The nanorobots were engineered to deliver anti-cancer drugs directly to the tumour cells, thereby minimising damage to the surrounding healthy tissue. This targeted drug delivery system could soon revolutionise cancer treatment in humans.

★ ★

Scientists have created a nanofibre membrane filter that can remove 99.9% of the salt from seawater in just minutes. If this technology can be scaled up, it could provide coastal communities with a virtually unlimited freshwater supply.

★ ★

Over the next few decades, nanoscale 3D printing could revolutionise the manufacturing process by enabling the creation of any three-dimensional object: from a diamond necklace to a fully functional television. This advanced technique would involve printing molecules or even individual atoms, then assembling them into larger, macro-sized objects. The advent of nanoscale 3D printing would transform fields like microelectronics, nanorobotics, battery technology, and drug design.

Environmental Technology

Some climate scientists have proposed combating climate change by dimming the Sun. One method to achieve this is to deploy high-altitude planes to release reflective particles into Earth's upper atmosphere. These particles would reflect sunlight back into space, mimicking the cooling effects of a large volcanic eruption. Research suggests that this strategy would be both cost-effective and feasible, with the potential to significantly reduce the effects of climate change. However, dimming the Sun would also have a range of negative consequences. These include increased acid rainfall, a delayed recovery of the ozone layer, and a disruption to monsoon rains that provide crops for hundreds of millions of people across Africa and Asia.

★ ★

Studies have found that if 100 planes were to release sulphur particles into the upper atmosphere regularly, the Sun's brightness would decrease by approximately 1%. This reduction in solar radiation would cause temperatures on Earth to drop by about one degree Celsius.

★ ★

Dimming the Sun would turn a clear blue sky into a milky white hue. This is because adding more reflective particles into the atmosphere would scatter sunlight more evenly, diminishing the dominance of blue light.

★ ★

Iron fertilisation is a proposed method to combat climate change that involves adding iron to regions of the ocean where it's scarcely found. This process would stimulate the growth of large phytoplankton blooms, which, upon dying, would sink to the ocean floor carrying the carbon dioxide they absorbed through photosynthesis. However, iron fertilisation has its drawbacks. It could worsen the problem of ocean acidification and promote the growth of unwanted phytoplankton species, potentially causing an imbalance in the ecosystem.

★ ★

Cloud seeding is a weather modification technique that involves spraying microscopic particles into clouds. These particles act as nuclei: encouraging the formation of ice crystals, attracting moisture, and eventually leading to rainfall. Cloud seeding is used to increase rainfall in many drought-prone countries, including the US, Russia, China, India, and the UAE. The US also employed it during the Vietnam War to extend the monsoon season and slow the Vietnamese military. Cloud seeding has been utilised for various other applications, including fighting forest fires, clearing fog from airports, and even diverting rainfall during major events like the 2008 Beijing Olympics.

★ ★

Several US cities—including Los Angeles, San Antonio, and Phoenix—have started painting their roads white to combat rising road and air temperatures. The white paint reflects the Sun's heat, reducing surface temperatures by up to nine degrees Celsius.

★ ★

Counterintuitively, a single-use plastic bag can have a lower environmental impact than a paper or cotton bag. This is because plastic bags are made from lightweight materials that require fewer resources to produce. In contrast, paper and cotton bags are more durable but demand significantly more energy and resources to manufacture. The key factor, however, is how often each bag is reused. According to a 2011 report, a paper bag would need to be reused three times, and an organic cotton bag 131 times, to match the low environmental footprint of a single-use plastic bag.

★ ★

The water treatment plant in Warsaw, Poland, uses eight clams to monitor the city's water supply. These clams are highly sensitive to pollutants, so when they detect contamination, they instinctively close their shells. This natural response triggers an alarm that automatically shuts off the water supply, preventing contaminated water from reaching the city.

Medical Technology

Scientists have created a device called BrainPort that enables blind people to 'see'. It uses a camera to capture visual information, which is then converted into electrical pulses. These pulses are sent to an array of electrodes placed on the tongue. The brain interprets these impulses as visual signals, processing them through the visual cortex to create a form of sight. Thanks to BrainPort, blind people can now perceive doorways, elevator buttons, obstacles, and even individual letters and numbers.

★ ★

In 2025, researchers developed a brain-computer interface that translates thoughts into speech nearly instantly. During trials, a woman with paralysis merely thought sentences in her mind. The device then accurately decoded the sentences and spoke them aloud using a synthetic voice.

★ ★

US neurotechnology company Neuralink has developed an implantable brain-computer interface that bridges the gap between the human brain and digital technology. The device, which is roughly the size of a coin, is surgically placed into the skull, where its ultra-thin wires monitor brain signals and convert them into computer commands. This breakthrough technology has allowed people with severe paralysis to operate external devices using their mind. In 2024, Noland Arbaugh, a 29-year-old quadriplegic, became one of the first people to test Neuralink's implant. Using just his thoughts, he successfully moved a computer cursor, adjusted music, and even played video games.

★ ★

CRISPR is a revolutionary technology that allows scientists to edit the genes of living organisms. Over the next few decades, it has the potential to cure thousands of genetic disorders, eradicate viruses like HIV and herpes, boost the body's ability to fight cancer, and even slow down the aging process. However, this powerful tool comes with ethical concerns. If CRISPR is used to eliminate genetic diseases in babies, it might also be exploited to provide babies with the genes for desirable traits, such as enhanced athleticism or intelligence. This could lead to irreversible changes to the human gene pool, potentially making genetically modified humans the new norm. Given these concerns, many scientists are calling for a ban on the use of CRISPR in human embryos.

★ ★

The magnets in some MRI machines are as powerful as those found in the Large Hadron Collider.

★ ★

An MRI machine uses powerful magnets and radio waves to create detailed images of inside your body. It works by using its strong magnetic field to align the protons in your body's water molecules. It then sends out radio waves to temporarily knock these protons out of alignment. When the protons return to their original positions, they emit faint signals that the MRI machine can detect and use to build precise imagery of your organs, muscles, and other internal structures.

★ ★

Positron emission tomography (PET) scans use antimatter to locate cancer in the body.

★ ★

Proton beam therapy is a cancer treatment that operates like a miniature particle accelerator. It works by focusing a narrow, high-energy beam of protons directly onto the tumour. This approach delivers radiation directly to the cancer cells, minimising exposure to the surrounding healthy tissue.

★ ★

Dermatologists have taken astronomy software that was originally designed to track changes in the sky and repurposed it to monitor high-risk melanoma patients. The patient's skin is photographed and treated as the background sky, while individual moles are treated as stars. The software enables dermatologists to closely track changes in moles over time, improving the early detection of melanomas.

★ ★

Germicidal UV lamps could soon become a standard feature in all public indoor spaces, reducing the spread of airborne diseases and preventing future pandemics. They work by emitting a specific type of ultraviolet light that has been proven to kill airborne viruses, bacteria, and other pathogens. However, the challenge with developing this technology is ensuring that the UV light is powerful enough to disinfect the air, while remaining safe for human skin and eyes.

Information

The weekend edition of *The New York Times* contains more information than the average 17th-century English person encountered in their lifetime.

★ ★

It's theoretically possible to encode the entire contents of every book ever written by simply marking a single notch on a metal rod. To achieve this, each letter needs to be assigned a numerical code: for example, A = 001, B = 002, C = 003, and so on, with 000 representing a space. Using this system, the text of every book ever written could be converted into a long string of digits. Placing a zero and a decimal point at the beginning of this number would turn it into a finite decimal fraction, which could be expressed as the ratio of two whole numbers. This ratio, in turn, could be represented by a single notch on a long metal rod. With the encoding key, someone could measure the notch's position, decode the ratio, and reconstruct the original text of every book ever written.

★ ★

DNA can store vast amounts of information in an incredibly small volume. In 2013, researchers demonstrated this potential by successfully encoding over five million bits of data—including all 154 Shakespearean sonnets, a photograph, and an audio clip—into DNA. The researchers later retrieved this stored data with 100% accuracy.

★ ★

A DNA storage system the size of a table tennis ball could hold 33 zettabytes of information. That's equivalent to all the data created, captured, copied, and consumed globally in 2018.

Laboratory

The Large Hadron Collider is so sensitive that the gravitational effects of the Moon must be taken into account when it collides protons.

★ ★

The magnets in the Large Hadron Collider collectively weigh five times more than the Eiffel Tower.

★ ★

The Large Hadron Collider produces up to one billion particle collisions every second. After just three days, these collisions create enough data to fill a stack of CDs as high as the Moon. However, since not all of these collisions are scientifically relevant, the Large Hadron Collider employs a sophisticated trigger system to filter out the uninteresting events in real time. This system narrows down the vast amount of data to just the 1,000 most promising collisions per second, ensuring that the researchers focus on only the most valuable results.

★ ★

A superconducting gravimeter is a scientific instrument that uses superconductors to measure tiny changes in Earth's gravitational field. Some superconducting gravimeters are so sensitive to changes in mass that they can tell if the scientists working on them have eaten lunch.

★ ★

The Laser Interferometer Gravitational-Wave Observatory (LIGO) is a scientific instrument that uses laser beams to measure tiny disturbances in space-time. It consists of two perpendicular arms, each four kilometres long, with mirrors suspended at either end. It is so precise that it can detect changes in the distance between its mirrors down to one ten-thousandth the width of a proton. To put this sensitivity into perspective, it's equivalent to measuring the 4.3-light-year distance from Earth to Proxima Centauri to an accuracy smaller than the width of a human hair. Due to this extreme sensitivity, something as slight as a person yawning from 20 metres away is enough to potentially affect its readings.

★ ★

Scientists can move matter using a beam of light—a technique known as optical tweezers. This method works because light carries momentum, enabling it to exert a force on microscopic objects and thereby move them. The discovery of optical tweezers earned Arthur Ashkin the 2018 Nobel Prize in Physics.

★ ★

Femto-photography is an advanced imaging technique that captures trillions of frames per second, allowing us to see the propagation of light itself. Using this technology, researchers have developed a camera that can see around corners, beyond the line of sight. The camera works by analysing the light reflected off walls to reconstruct a detailed, three-dimensional image of the hidden object. This technology opens up new possibilities in medical imaging, driverless vehicles, and search-and-rescue operations.

★ ★

In 2018, scientists developed a method to shrink objects to the nanoscale. Their technique involves embedding materials in gel and exposing them to laser light—a process that reduces simple structures to one-thousandth of their original size. This breakthrough paves the way for the creation of highly detailed, miniature structures that have potential applications in medicine, electronics, and robotics.

Historical Inventions

Ancient Roman concrete was superior to modern-day concrete. In 2023, scientists reverse-engineered its formula to uncover the secret to its strength. They discovered that the Romans mixed a mineral called quicklime with seawater, creating a self-repairing concrete capable of lasting thousands of years. To put this longevity into perspective, modern-day concrete can begin to degrade in as few as 50 years.

★ ★

In 1901, an ancient, hand-powered model of the Solar System was discovered off the coast of the Greek Island of Antikythera. This device, named the Antikythera mechanism, is considered the world's oldest analogue computer. Created around the 2nd century BCE, it features a highly intricate system of 37 bronze gears and a complex arrangement of dials. It was so advanced for its time that it could track planetary positions to an accuracy of one degree over 500 years, monitor the movement of the Moon and stars, and even predict eclipses decades in advance.

★ ★

During the 19th century, electric cars could reach speeds of over 100 kilometres per hour.

★ ★

Early wireless TV remotes used sound to change the channel. When a button was pressed, a small hammer inside the remote would strike an aluminium bar, producing an ultrasonic sound that the TV could detect. However, this technology had a significant flaw: everyday household noises, like opening a soft drink can or clinking two spoons together, would accidentally change the channel.

★ ★

During the 1980s, video games could be downloaded directly from the radio. Radio stations would broadcast data over the airwaves, which listeners could record on cassette tapes. These tapes could then be loaded onto a home computer, turning the audio into a playable video game.

Modern Inventions

Google has developed augmented reality glasses that can instantly translate speech from one language to another and display the translated text in front of the user's eyes. The glasses provide real-time subtitles for the world around you.

★ ★

In 2006, Japanese scientists created an 'odour recorder' that can store, analyse, and reproduce smells. Using an array of sensors, the device detects the chemical composition of a scent, stores the data, and then recreates the aroma by mixing various chemical compounds.

★ ★

In 2012, Japanese researchers developed a 'speech-jammer gun' that can stop a person from speaking when it's pointed at them. The device works by recording a person's voice and playing it back to them with a slight delay. This disrupts normal speech by interfering with auditory feedback—the brain's reliance on hearing its own voice in real time to maintain smooth and coherent speech.

★ ★

Wireless power transfer is a technology that enables electrical devices to receive power without the need for physical wires. Unlike wireless charging, which requires a device to be in direct contact with a charger, wireless power transfer uses electromagnetic fields to transmit electricity through the air. This technology can deliver electricity to multiple devices simultaneously across an entire room, much like how Wi-Fi provides internet connectivity. Wireless power transfer has already been adopted by some homes and businesses and will likely be used universally in the near future.

★ ★

Li-Fi is a wireless communication technology that transmits data using light rather than radio frequencies. By rapidly modulating light signals, it can achieve speeds 100 times faster than Wi-Fi. However, to work efficiently, it requires a direct line of sight between the light source and the device, making it less versatile than Wi-Fi.

★ ★

In 2025, China unveiled a powerful laser camera capable of capturing distant images with extraordinary precision. During tests conducted in ideal conditions, the camera achieved millimetre-level clarity, successfully detecting objects as small as 1.7 millimetres from 100 kilometres away. With this degree of resolution, the camera could theoretically remain stationed on Earth and be able to read the serial numbers of satellites in orbit.

★ ★

Researchers are developing a 3D printer that can print itself. They have already created a 3D printer that can print most of its own parts and are now working towards a 100% replication.

Miscellaneous Technology

The 'white' you see on your television, smartphone, and computer screen isn't actually white—it's an optical illusion. Each pixel on your screen is made up of three tiny subpixels: red, green, and blue. When these three subpixels are illuminated simultaneously and viewed from a distance, their colours blend together, tricking your eyes into perceiving white. To see this phenomenon for yourself, use a magnifying glass to look closely at the white parts of your screen. You won't see any white at all; instead, you'll see the individual red, green, and blue dots that create the illusion.

★ ★

Using only a balloon, you can make an unplugged fluorescent light bulb glow. To achieve this, rub the balloon against your hair or clothing to build up static electricity. Next, in a dark room, wave the balloon above the unplugged light bulb. The static charge will cause the bulb to light up briefly.

★ ★

Surgical light doesn't cast a visible shadow.

★ ★

To increase the range of your car key remote, try holding it against your temple when pressing the button. The radio waves from the remote will pass through your water-dense body, effectively turning you into a dielectric resonator antenna. This will increase the amplitude of the waves, potentially doubling the range of the signal.

★ ★

Dead batteries bounce higher than fully charged batteries.

★ ★

Recycling scrap aluminium uses just 5% of the energy required to produce new aluminium. Thanks to this remarkable efficiency, 75% of all the aluminium ever produced is still in use today.

★ ★

SpiKey technology can create a physical copy of a key by analysing the sounds it makes when it's inserted into a lock. When a key turns, it produces unique 'clicks' caused by its distinctive shape interacting with the lock's internal pins. SpiKey uses these audio cues to reverse-engineer the key's design and generate a duplicate.

★ ★

During the 1930s, Ohio's WLW radio station was the most powerful commercial station in the world, broadcasting at half a million watts. At night, its AM radio waves interacted with Earth's ionosphere, transforming into skywaves that could be heard across much of the US. The station's immense power led to some unusual phenomena: a nearby service station couldn't turn its lights off; gutters rattled loose from adjacent buildings; and locals reported hearing the radio broadcast in everyday objects, such as pots, pans, and mattresses.

★ ★

If the entire internet were printed out on A4 sheets of paper, the stack would be taller than Earth's diameter.

★ ★

The year 2038 problem presents a potentially greater challenge than the Y2K problem. It has arisen because many computers track time by counting the number of seconds that have elapsed since 1 January 1970. These systems rely on 32-bit integers, which can only represent values up to $2^{31} - 1$ seconds. As a result, the last time they can encode is 03:14:07 UTC on 19 January 2038. Beyond this time, an integer overflow will occur, potentially causing them to malfunction. To address this issue, some modern systems have been upgraded to measure time in 64-bit integers. These systems can represent time for 292 billion years—about 21 times the current age of the Universe—before encountering an integer overflow.

★ ★

Colossal Biosciences is a US biotechnology and genetic engineering company that aims to use advanced genetic techniques to bring extinct species back to life. The company is currently working to revive several species, including the woolly mammoth, the dodo bird, and the Tasmanian tiger.

★ ★

In 2025, Google unveiled DolphinGemma, an artificial intelligence model that can interpret dolphin sounds and respond with its own dolphin-like vocalisations. This technology could pave the way for interspecies communication between humans and other intelligent animals in the near future.

★ ★

Xenobots are programmable life forms, about the size of a grain of sand, made from frog stem cells. They can walk, swim, carry payloads, clean up debris, work together in groups, and even self-replicate to create new xenobots. Scientists remain divided over whether xenobots should be classified as organisms, robots, or something entirely different.

ZOOLOGY

Fish

The modern classification system groups animals, plants, and other organisms based on their evolutionary history. However, since all vertebrates evolved from ancient water-dwelling creatures that we call 'fish', this presents a dilemma. Either every land vertebrate—including humans, birds, and dogs—must be classified as fish, or the category 'fish' doesn't truly exist in taxonomy.

★ ★

The climbing perch is a freshwater fish that crawls on land and climbs trees using its strong pectoral fins. It can survive out of water for up to six days thanks to a specialised organ called the labyrinth, which allows it to breathe air.

★ ★

Dead fish can 'swim' upstream, against the flow of oncoming water. This occurs when the water flowing around a fish's body creates alternating vortices that push the fish against the direction of the current.

★ ★

Freshwater lakes and rivers account for just 0.01% of Earth's water, yet they support half of all fish species.

★ ★

Freshwater fish urinate nonstop. The body of a freshwater fish is saltier than the surrounding water, so water naturally seeps into its body through osmosis. To remove this excess water and keep its internal salt levels stable, a freshwater fish produces a large amount of very dilute urine, which it continuously expels.

★ ★

Several species of fish have been observed using tools to obtain food. For example, they will grasp a sea urchin or clam in their mouth and repeatedly strike it against a rock until the shell breaks. This allows them to feed on the soft tissue inside the shell.

★ ★

All clownfish are born male. When the female in a group dies, the dominant male undergoes an irreversible sex change to become female. This process is known as sequential hermaphroditism.

★ ★

The sailfish is the fastest fish in the world, with a top speed of 110 kilometres per hour. To put this speed into perspective, it could cover the length of an Olympic-sized swimming pool in just 1.6 seconds.

★ ★

The dwarf seahorse is the slowest fish in the world, with a maximum speed of just 1.5 metres per hour. At this pace, it would take 33 hours to travel the length of an Olympic-sized swimming pool.

★ ★

Seahorses—along with their sea dragon and pipefish relatives—are the only species in which the male becomes pregnant and gives birth. This process begins when the female seahorse deposits up to 1,500 eggs into the male's pouch, where he fertilises them. As the embryos develop, the male's abdomen expands, much like in human pregnancy. When it's time to give birth, the male's pouch opens, releasing the young seahorses into the water.

★ ★

Scientists have discovered a species of ultra-black fish whose skin absorbs 99.95% of light. This adaptation makes its body 200 times darker than black paper, rendering it virtually invisible in the depths of the ocean.

★ ★

Lungfish are more closely related to humans than to salmon.

★ ★

Some species of fish navigate, communicate, and hunt using electrical signals. When two fish with overlapping frequencies encounter each other, one shifts its frequency upwards, while the other shifts its frequency downwards. This adjustment minimises the frequency overlap, allowing both fish to continue using their electrical signals without interference.

★ ★

The giant trevally fish eats birds. It typically targets birds resting on the ocean's surface, but in some cases, it will leap out of the ocean to catch a bird in mid-flight.

★ ★

Flying fish can leap out of the water and glide through the air for considerable distances. In fact, some species can stay airborne for up to 45 seconds, travelling as far as 400 metres at speeds of up to 70 kilometres per hour. Although gliding through the air helps flying fish evade underwater predators, it comes at a cost—it makes them vulnerable to attacks from birds.

★ ★

Some whipnose anglerfish spend their entire life swimming upside down.

★ ★

Zebra mbuna fish can solve simple addition and subtraction problems.

★ ★

Grouper fish and giant moray eels work together to hunt prey more effectively. The grouper initiates the hunt by approaching the eel and signalling its readiness with a vigorous head shake. Once the signal is given, the grouper and eel launch a coordinated attack, leveraging their complementary hunting styles. The eel uses its slender body to flush out prey hiding in cracks and crevasses, while the grouper chases down prey that escapes into open water. This teamwork allows both predators to hunt more successfully than they could alone.

★ ★

The cleaner wrasse will swim directly into a shark's mouth to feed on parasites and leftover food. The shark understands that this is beneficial to its health, so it refrains from eating the wrasse. This unique symbiotic relationship provides the wrasse with a meal, while keeping the shark clean and healthy.

Sharks

The Greenland shark has the longest lifespan of any known vertebrate. In fact, a 2016 study found that one particular Greenland shark had lived for 392 years, with a margin of error of 120 years. This means it was born sometime between 1504 and 1744.

★ ★

Female Greenland sharks can only reproduce once they reach about 150 years of age.

★ ★

Female Greenland sharks are believed to have the longest gestation period of any animal, estimated at between 8 and 18 years.

★ ★

Nearly all Greenland sharks eventually lose their sight. They are born with small, yet functional, eyes. However, over time, parasitic crustaceans gradually eat away at their corneal tissue, making them blind. Despite their disability, Greenland sharks still thrive in their deep-sea environment because they rely on other senses—such as smell and electroreception—to navigate and locate their prey.

★ ★

The female sand tiger shark has two wombs. After impregnation, each womb holds between 16 and 23 fertilised eggs. When the most developed shark embryo in each womb grows teeth and hatches, it eats all of the siblings that it had been sharing a womb with—a process called intrauterine cannibalism. This ensures that just one pup from each womb survives to reach the open sea.

★ ★

When a shark is flipped onto its back, it experiences a reflex called tonic immobility, which puts it into a subdued, hypnotic state, rendering it temporarily paralysed. Orcas have been observed taking advantage of this reflex to hunt sharks. They will hold a shark upside down, inducing tonic immobility, thereby leaving the shark defenceless.

★ ★

Sharks are older than dinosaurs, trees, the Atlantic Ocean, the rings of Saturn, and even the North Star.

Whales

The volume of water that a blue whale can hold in its mouth can weigh as much as the blue whale itself.

★ ★

A blue whale's tongue can weigh as much as an entire elephant.

★ ★

The arteries of a blue whale are so large that a bowling ball could pass through them.

★ ★

A blue whale's heart rate can drop to as low as two beats per minute during a deep dive.

★ ★

A blue whale's throat is so narrow that it can't swallow anything larger than a grapefruit.

★ ★

Sperm whales use echolocation to track and hunt their prey. Their echolocation clicks are the strongest biosonar signals found in nature. They are so powerful that they can detect something as small as a 25-centimetre-long squid from up to 1.7 kilometres away.

★ ★

Sperm whale clicks are the loudest biologically generated sound on Earth, reaching volumes of up to 230 decibels.

★ ★

Sperm whale clicks are so loud that they can potentially harm humans. In one documented incident, a diver's hand was paralysed for four hours after being exposed to a sperm whale's clicks. There are additional reports of sperm whale clicks causing divers' bodies to heat up.

★ ★

Seagulls have been observed preying on southern right whales off the coast of Argentina. When a whale surfaces to breathe, the seagulls swoop down and take bites out of the whale's skin and blubber. In response to these attacks, the local whales have recently developed a new breathing technique called oblique breathing. This involves raising their blowhole just enough above the water's surface to breathe, but not enough to expose their flesh to the seagulls.

★ ★

A fin whale's call can travel more than 6,000 kilometres—a distance equivalent to crossing the Atlantic Ocean, from New York State to Morocco.

★ ★

Orcas have been observed adopting fashion trends. In 1987, a female orca started a unique trend by wearing a dead salmon on her nose, much like a hat. Her behaviour quickly spread. Soon after, other orcas in her pod—and even those in two neighbouring pods—were also seen wearing a dead salmon on their nose.

★ ★

The bowhead whale is the longest-living mammal on Earth, with a lifespan of over 200 years. Incredibly, some bowhead whales swimming in our oceans today were born before Herman Melville wrote *Moby-Dick*.

★ ★

Whales and fish have distinct swimming styles due to their different evolutionary backgrounds. Whales swim by moving their tail up and down, a trait inherited from their land-dwelling ancestors, whose spines flexed vertically as they ran. This same up-and-down motion can still be seen in modern mammals, such as dogs, when they run. In contrast, fish swim by moving their tail from side to side, reflecting their distinct aquatic evolution.

Octopuses

When an octopus needs to cross an exposed area, it will mimic an inanimate object to avoid detection. If it chooses to mimic a plant, it waves some of its arms to make them look like swaying branches, while using its other arms to sneak forwards. In rough waters, it moves quickly, much like how a plant would in turbulent conditions. However, in calm waters, it acts as though it's rooted to one spot, while slowly inching forwards—hoping its predators don't notice it's moving. This stealthy strategy can turn what would normally be a 20-second crossing into a 20-minute journey.

★ ★

Octopuses have nine brains, three hearts, and blue blood.

★ ★

When an octopus loses an arm, that arm can remain functional for up to one hour, despite being detached. In some instances, the severed arm will even attempt to pick up food and bring it to where the mouth would have been, as if it were feeding a phantom body.

★ ★

The only hard part of an octopus's body is its beak, which is relatively small. As a result, a medium-sized octopus can squeeze its entire body through a hole the size of a US quarter coin.

★ ★

Female and male blanket octopuses exhibit one of the most extreme size differentials in the animal kingdom. Females grow to about two metres in length, while males are significantly smaller, growing to just 2.4 centimetres. This size disparity is accompanied by an equally remarkable weight difference, with some females weighing 40,000 times more than their male mates.

★ ★

In 2018, a team of 33 scientists published a paper proposing a controversial hypothesis: octopuses are in fact alien. According to this theory, octopuses arrived on Earth as frozen octopus eggs carried by a meteor, or as the result of an alien virus infecting a primitive population of squid. Supporters of the octopus-alien hypothesis point to the fact that the octopus genome is highly complex, containing roughly 33,000 protein-coding genes—significantly more than the 20,500 found in humans. They also highlight the octopus's remarkable physical traits, including its uniquely large primary brain, sophisticated central nervous system, incredibly flexible body, camera-like eyes, and ability to change colour and shape instantly. However, despite these factors, the octopus-alien hypothesis remains fringe and unproven—the overwhelming consensus among zoologists is that octopuses evolved on Earth through standard evolutionary processes.

Shrimp

The pistol shrimp employs a unique strategy to hunt its prey—it snaps its claw shut so fast that it creates a vacuum in the water, known as a cavitation bubble. This bubble shoots away from its claw at 25 metres per second. When it collapses, it generates a brief flash of light, a phenomenon known as sonoluminescence. During this collapse, the bubble heats up to 4,800 degrees Celsius—comparable to the surface temperature of the Sun. The pistol shrimp uses the shock wave from this collapsing bubble to stun and kill its prey.

★ ★

When the pistol shrimp snaps its claw shut, it generates a sound of 218 decibels. This makes it one of the loudest animals on Earth, alongside sperm and blue whales. Pistol shrimp colonies are so loud that they are a major source of oceanic noise, disrupting sonar systems, underwater communications, and anti-submarine warfare.

★ ★

Juvenile pistol shrimp can snap their claw shut with an acceleration of 580,000 metres per second squared. That's equivalent to the acceleration of a bullet exiting the barrel of a gun.

★ ★

A mantis shrimp's strike is so powerful that it can break a glass aquarium. To prevent this from happening, many researchers keep their mantis shrimp in plastic tanks.

★ ★

The mantis shrimp has the most mobile eyes in the animal kingdom. While the majority of animals minimise their eye movement to prevent motion blur, the mantis shrimp defies the norm by constantly moving its eyes. This continuous scanning allows it to rapidly gather visual information so that it can detect predators and prey.

★ ★

The mantis shrimp has trinocular vision, enabling it to see objects through three distinct regions of the same eye. It also possesses 16 colour receptors—compared to the three found in humans—and can perceive ultraviolet and polarised light.

Aquatic Animals

Dolphins can 'see' through objects using echolocation. By producing high-frequency clicks and analysing the returning echoes, they can identify shapes hidden behind an opaque barrier, determine the contents of other dolphins' stomachs, and even potentially detect human pregnancies.

★ ★

Dolphins communicate using unique 'signature whistles' that function as names. These whistles are one of the most advanced forms of animal communication outside of human language.

★ ★

Limpets are a type of aquatic snail renowned for using their strong, muscular foot to cling tightly to surfaces. Their teeth are the toughest biological material known to exist, with a strength five times that of spider silk (previously considered the toughest biological material). Limpet teeth are even tougher than most human-made materials, including Kevlar. In fact, their toughness rivals that of the most advanced carbon fibre composites.

★ ★

The glass sponge can live for up to 15,000 years.

★ ★

Whenever a hydra—a freshwater polyp closely related to the jellyfish—is ready to eat, it rips itself open a new mouth opening. Once the meal is finished, it stitches up the mouth hole.

★ ★

The hydra is theoretically immortal. It's primarily made up of stem cells that continually regenerate. This prevents it from aging.

★ ★

The jellyfish Turritopsis dohrnii is biologically immortal. When physically damaged, it reverts back to an earlier developmental stage called a polyp, which is genetically identical to its adult form. While Turritopsis dohrnii can still die if eaten by a predator, it has the potential to live indefinitely under the right conditions. In fact, it's theoretically possible that a Turritopsis dohrnii has been floating in the ocean for 66 million years—the entire duration of the species' existence.

★ ★

A jellyfish is composed of 95% water and just 5% solid material. Therefore, when it's left in the hot Sun, most of its body will evaporate.

★ ★

In 1991, NASA sent 2,478 jellyfish polyps into space to study how microgravity would affect the development of their offspring. Over nine days, many of the jellyfish reproduced, resulting in a total of 60,000 jellyfish orbiting Earth by the mission's end. When the newly-hatched jellyfish returned to Earth, they struggled to distinguish between up and down because their gravity receptors didn't develop properly in space. Since humans have similar gravity receptors, this suggests that if human babies are ever born in space, they might face the same difficulties in adapting to Earth's gravity.

★ ★

The lion's mane jelly is the world's longest species of jellyfish. It can grow up to 36.5 metres in length—longer than a blue whale.

★ ★

The comb jelly can devour prey that's larger than its own body. It achieves this by stretching its flexible, transparent body around its prey to fully engulf it.

★ ★

When a comb jelly is injured, it can fuse with another comb jelly to form a 'single organism' with a shared nervous system.

★ ★

The fastest-growing multicellular animal on Earth is a gelatinous marine creature called a salp. It can increase its body length by 10% per hour and grow to maturity in just two days.

★ ★

The squid's brain is shaped like a doughnut with a hole in the centre. Because its oesophagus passes directly through this central opening, if it eats something too large, it risks getting brain damage.

★ ★

The colossal squid has the largest eyes in the animal kingdom. Each eye measures about 27 centimetres in diameter—the size of a dinner plate.

★ ★

Carcinisation is the phenomenon in which different groups of animals continue to independently evolve into crabs. This evolutionary process has occurred on at least five separate occasions, suggesting that natural selection favours the crab design.

★ ★

The nautilus is a marine mollusc that's often referred to as a 'living fossil' because it has remained virtually unchanged for 500 million years. It essentially evolved into its current form early in evolutionary history and has stayed that way ever since.

★ ★

A zooid is a general term for an individual organism, usually just a few millimetres long, forming part of a larger colony. When millions join together, they create a hollow, tube-shaped superorganism called a pyrosome, which can be as large as a sperm whale. This collective structure allows the zooids to capture plankton, share nutrients, and defend against predators. In many pyrosomes, zooids share tissues, fluids, and even nerve signals.

★ ★

Antarctic krill are roughly the size of your pinkie finger. Yet, when they gather together, they can form massive super-swarms—tens of kilometres wide—visible from space.

★ ★

Chalk is made up of trillions of ancient, microscopic plankton skeletons.

★ ★

Russian cosmonauts have discovered sea plankton living on the outside of the International Space Station. Scientists believe these microscopic organisms may have been carried up there by air currents.

Snakes

The inland taipan is the world's most venomous snake—a single bite contains enough venom to kill 100 people. Despite this extreme potency, no human fatalities from its bite have ever been recorded.

★ ★

When threatened by a predator, certain snake species employ a dramatic defensive mechanism known as autohaemorrhaging. This involves intentionally raising the blood pressure in their head so that blood seeps from their eyes, mouth, and nostrils. By creating such a graphic spectacle, the predator typically becomes startled, giving the snake a critical opportunity to escape.

★ ★

When threatened, the eastern hognose snake 'plays dead' by rolling onto its back and lying completely still. To make the act even more convincing, it sometimes releases a foul odour and lets its tongue hang out.

★ ★

The pit viper hunts for prey at night using the powerful heat-sensing organs on either side of its head. These organs are so sensitive to heat that they can detect temperature variations of just 0.003 degrees Celsius—the equivalent of noticing if a single drop of hot water were added to a swimming pool. They provide a hunting advantage by allowing the pit viper to perceive the thermal radiation emitted by warm-blooded animals in complete darkness.

★ ★

A snake's eyes are technically always closed. Instead of eyelids, snakes have a transparent layer of skin called a spectacle that permanently covers their eyes. This built-in shield keeps their eyes safe, while still allowing them to see clearly.

Reptiles

Crocodiles will climb up to five metres into a tree. They do this to bask in the sunlight, regulate their body temperature, and watch for predators and prey.

★ ★

The saltwater crocodile has the strongest confirmed bite of any living animal. Its jaws can exert a crushing force of 3,700 pounds per square inch—equivalent to 23 times the average human bite strength. Yet, despite having this incredible biting power, a crocodile's jaws are extremely weak when it comes to opening. In fact, they can be held shut with something as simple as a rubber band.

★ ★

The male alligator has a permanently erect penis tucked inside its body. When it's time to mate, it pops out, and once the process is complete, it retracts back inside its body.

★ ★

The Komodo dragon can taste its prey from up to four kilometres away. It achieves this by using the tips of its forked tongue to collect scent particles from the air. By analysing which side of its tongue has more particles—left or right—it can even determine the direction of its prey.

★ ★

The Labord's chameleon spends most of its life—up to nine months—developing inside an egg. Once it hatches, it only lives for another four to five months.

★ ★

The gender of a sea turtle hatchling is determined by the sand temperature in which its egg incubates. If the sand temperature is below 27.7 degrees Celsius, the hatchling will be male, whereas if the temperature is above 31 degrees Celsius, the hatchling will be female. Due to the recent rise in global temperatures, the natural gender ratio of sea turtles is becoming increasingly skewed. For instance, on Raine Island, Australia, female sea turtles now outnumber males by a ratio of 116 to 1.

Birds

Birds are classified within the clade Dinosauria, which means they are not just descendants of dinosaurs—they actually are dinosaurs. They evolved from theropods, a group of two-legged dinosaurs that included species like Tyrannosaurus rex and Velociraptor. Over millions of years, some theropods developed feathers, smaller bodies, and the ability to fly, eventually becoming the birds we see today.

★ ★

The bee hummingbird is the smallest species of bird and the tiniest known dinosaur. It's so small—about the length of your pinkie finger—that people often mistake it for an insect.

★ ★

Hummingbirds can fly upside down and backwards. This is because, unlike most birds, they have a ball-and-socket joint at the shoulder that allows their wings to rotate 180 degrees in all directions.

★ ★

A hummingbird's metabolism is so fast that it must feed every 15 minutes and consume half of its body weight in food each day. If humans had a similar metabolism, we would need to eat over 150,000 daily calories just to survive. Due to its extreme metabolic rate, the hummingbird is always just a few hours away from starving to death. To survive through the night without eating, it enters a hibernation-like state called torpor, which conserves its energy until it can feed again.

★ ★

If a bald eagle sheds a feather on one wing, it will simultaneously shed the same feather on its other wing to maintain balanced flight.

★ ★

The bearded vulture is the only known animal whose diet consists almost exclusively of bone. When it encounters a bone that's too large to swallow, it carries it to a great height and drops it onto the rocky terrain below, thereby shattering it into smaller, more manageable pieces. The bearded vulture can digest bone because its stomach produces a substance that's almost pure hydrochloric acid, with a pH of around 0.7.

★ ★

The peregrine falcon is the world's fastest animal. It has a highest recorded diving speed of 389 kilometres per hour—equivalent to the fastest speed ever obtained by a Formula 1 car. At this incredible pace, it covers 270 of its body lengths every second.

★ ★

When the peregrine falcon executes a high-speed dive, the force of air rushing into its nostrils should, in theory, make breathing impossible. However, it has evolved tiny, bony tubercles inside its nostrils that channel the incoming air into a spiral path, allowing it to breathe effortlessly, even at incredible speeds.

★ ★

Owls can only look straight ahead because their eyes are fixed in their sockets. To overcome this limitation, they bob their head up and down, so each eye gains a slightly different angle of the scene. This creates a difference, known as parallax, which enables their brain to calculate depth and distance, giving them a three-dimensional view of the world.

★ ★

The common swift spends up to ten months continuously in the air without landing. It eats, sleeps, and even mates while flying.

★ ★

The Arctic tern holds the record for the longest migration of any animal. Every year, it flies from the Arctic to the Antarctic and back. Over its 30-year lifespan, this journey adds up to 2.1 million kilometres—the equivalent of flying to the Moon and back twice.

★ ★

While nesting, the chinstrap penguin sleeps more than 10,000 times a day. Each sleep episode lasts about four seconds, but this accumulates to 11 hours of daily sleep. This unique sleeping pattern allows the chinstrap penguin to keep an eye on its nest and stay vigilant against predators.

★ ★

Prostitution has been documented in Adelie penguins. Female Adelie penguins require small stones to build a nest for their eggs. However, since these stones are often in short supply along the beach, some females will mate with a single male in exchange for a stone.

★ ★

Some penguin species exit the water by jumping three metres into the air. They accomplish this feat by trapping a layer of bubbles around their body, which reduces drag and provides a cushion of air. This allows them to propel their body upwards with incredible force. Engineers have adopted this same technique to improve the efficiency of ships and torpedoes.

★ ★

Pigeons have been trained to recognise all 26 letters of the English alphabet and to distinguish real words from non-words.

★ ★

Pigeons have been trained through operant conditioning to differentiate 'good' art from 'bad' art.

★ ★

In some parts of China, fishermen use cormorants to help them catch fish. For every seven fish that the cormorant catches, it's rewarded with one fish to eat for itself. If the cormorant isn't given its share, it stops working. This behaviour suggests that cormorants can count up to seven.

★ ★

Firehawks are a group of Australian birds that have developed a unique hunting strategy. They deliberately spread fires by picking up burning sticks and dropping them in unburnt areas. Once the flames take hold, they wait nearby until they see small animals attempting to escape the blaze. At that point, they swoop in to make their kill.

★ ★

The Australian white ibis has developed a clever strategy to safely eat poisonous cane toads. It picks up a toad, tosses it around, and then rinses it in a water source. The tossing action causes the toad to release its poison, while the washing process removes this poison, making the toad safe to eat.

★ ★

Crows can remember a human face they've associated with a bad experience for up to five years.

★ ★

Flamingos are born with grey feathers. Over time, the carotenoid pigments from their diet turn their feathers into the vibrant pink colour that they're famous for.

★ ★

The western grebe is the heaviest bird capable of walking on water. Despite growing up to 75 centimetres tall and weighing as much as two kilograms, it can walk on water for seven seconds, covering a distance of up to 20 metres. While the western grebe excels at walking on water, it struggles to walk on land due to its unique anatomy.

★ ★

During the 1940s, Mike the chicken survived for 18 months after his head was chopped off. He retained the ability to stand, walk, and perch because most of his brainstem remained intact. His owners kept him alive by dripping water and liquid food directly into his oesophagus.

★ ★

For decades, one of the greatest mysteries in zoology has been 'how do birds navigate using Earth's magnetic field?' Recent research points to an astonishing explanation: birds navigate using quantum mechanical processes in their eyes. Within a bird's eyes, light-sensitive molecules known as cryptochromes detect magnetic fields. When light strikes these molecules, it creates a pair of particles with a quantum property called 'spin'. The interactions between these spins and Earth's magnetic field help the bird determine direction, like a built-in compass.

Bees

A bee has such an acute sense of smell that it can detect trace amounts of drugs, confirm human pregnancies, monitor fertility cycles, and even identify diseases such as skin cancer, lung cancer, diabetes, and tuberculosis.

★ ★

Bees understand the concept of zero.

★ ★

Bees are capable of flying at altitudes higher than Mount Everest.

★ ★

To produce just one kilogram of honey, bees from a single hive must visit four million flowers and fly a distance equivalent to four times around the Earth's circumference.

★ ★

Out of the 100 crop species that provide 90% of the world's food, 71 rely on bees for pollination. Consequently, if bees were to ever go extinct, human diets would need to change drastically.

★ ★

One quarter of all known bee species haven't been sighted since the 1990s.

★ ★

When a bee discovers a rich source of pollen, it communicates this information to the rest of its hive by performing a unique series of movements called the waggle dance. The duration of the dance reveals how far the pollen is from the hive, the liveliness of the dance indicates the quality of the pollen, and the angle of the dance signals the direction of the pollen source relative to the position of the Sun.

★ ★

A bee flapping its wings generates a positive electric charge, while flowers typically carry a slight negative charge. Due to this charge differential, a bee's antennae and tiny hairs vibrate when it approaches a flower. If the bee doesn't feel these vibrations, it indicates that another pollinator has recently visited the flower. If this is the case, the bee knows to move on and search for food elsewhere, thereby saving time and energy.

Ants

According to a 2022 study, there are at least 20 quadrillion ants on Earth. That equates to roughly 2.5 million ants for every person.

★ ★

All the ants on Earth weigh more than all the wild birds and mammals combined.

★ ★

There is a single supercolony of ants spanning 6,000 kilometres along the Mediterranean coastline, from northern Italy to the Atlantic coast of Spain.

★ ★

Ants have been observed self-medicating. When an ant contracts a dangerous fungal disease, it fights the disease by consuming food rich in hydrogen peroxide. This significantly boosts the chance of its survival.

★ ★

Several ant species farm and milk aphids, much like how humans farm and milk cows. The ants start this process by herding the aphids around the juiciest parts of a plant, where the aphids feed on sap. Once the aphids have consumed the sap, the ants 'milk' them for a sugary by-product called honeydew. In exchange for this food source, the ants protect the aphids from predators. Some ants even take this relationship a step further—during the winter, they carry aphid eggs into their nest to protect them until they hatch, ensuring a steady supply of honeydew all year long.

★ ★

If an ant sustains an injury to its upper leg, its nestmates will gnaw at the leg to amputate it. This behaviour is the only known example of a non-human species performing a life-saving amputation.

★ ★

Army ants are completely blind. They navigate by leaving behind a pheromone trail, which other army ants follow. However, this system can sometimes go awry when the pheromone trail forms a loop known as a 'death spiral'. It's named as such because army ants will follow each other in a circular path until they die of exhaustion.

★ ★

Some species of worker ants in Southeast Asia are known as exploding ants because, when threatened, they intentionally contract their abdominal muscles so forcefully that they cause their own body to rupture. This releases a toxic substance that can immobilise or even kill the attacker. While this act of self-sacrifice, known as autothysis, results in the ant's death, it ultimately benefits the colony by protecting the other ants.

★ ★

In 2006, researchers conducted an experiment to determine how ants measure distance. They divided the ants into two groups: one group had stilt-like extensions attached to their legs to lengthen their stride, while the other group had their lower legs shortened to reduce their stride. The ants were then observed returning to their nest. The ants with stilts overshot their nest, while the ants with stump-legs fell short of it. Based on these observations, the researchers concluded that ants use an internal pedometer to count their steps and gauge distance.

Insects

The world's smallest adult insect is the parasitic wasp Dicopomorpha echmepterygis, which measures just 0.127 millimetres in length. To put its size into perspective, if four of these wasps were lined up end to end, they would span the full stop at the end of this sentence. Due to its tiny size, this wasp doesn't fly in the traditional sense but instead drifts through the air on currents, much like a speck of dust.

★ ★

The Chinese stick insect Phryganistria chinensis is the world's longest insect, measuring up to 64 centimetres in total length. Despite its record-breaking size, it wasn't discovered until 2014.

★ ★

The physical traits of a female fly's offspring can be influenced by her previous sexual partners. For instance, if she mates with a large male fly and later has an offspring with a small male fly, the offspring may exhibit the larger stature of the first sexual partner. This phenomenon, known as non-genetic telegony, is thought to occur because molecules from the semen of earlier partners are absorbed into the female's immature eggs, altering the traits of her future offspring.

★ ★

Flies don't have teeth, so they cannot take bites out of solid food. To eat, they vomit digestive juices onto their food, so it becomes a liquefied soup that they can easily swallow.

★ ★

The female Dolania americana mayfly has the shortest adult lifespan of any known animal. After spending up to two years as an aquatic nymph on the riverbed, it eventually emerges as a flying adult—only to live for less than five minutes. During this brief period, it mates, lays eggs, and dies.

★ ★

Queen termites can live for up to 50 years.

★ ★

Queen termites are incredibly prolific egg-layers, capable of laying one egg every three seconds. This adds up to 30,000 eggs per day and 164 million eggs over their lifetime.

★ ★

Northeastern Brazil is home to a vast network of 200 million termite mounds that cover an area the size of Great Britain. Some of these mounds are up to 4,000 years old—comparable in age to the Great Pyramid of Giza. Remarkably, they are still inhabited by the same species of termite that originally constructed them.

★ ★

When a raindrop hits a mosquito, the force is comparable to a car falling on a person. Yet, mosquitoes have developed a clever strategy to survive these collisions—briefly riding on the falling raindrop, then detaching before it hits the ground.

★ ★

A male silkworm moth can smell a female up to 11 kilometres away.

★ ★

The dung beetle can pull an object that's over 1,000 times its own body weight.

★ ★

The nocturnal dung beetle, Scarabaeus satyrus, uses the Milky Way to navigate. In a 2013 study, scientists confirmed this by placing dung beetles in a planetarium and manipulating the artificial light so that only the Milky Way was visible. The dung beetles successfully orientated themselves using just the artificial Milky Way.

★ ★

In 2014, it was announced that Charles Darwin had discovered a new species of beetle. Darwin collected the beetle in Argentina in 1832, and it was stored in London's Natural History Museum. The specimen remained unnoticed until 2008, when entomologist

Stylianos Chatzimanolis rediscovered it while combing through historical records. Chatzimanolis named the beetle Darwinilus sedarisi in honour of Charles Darwin.

★ ★

Beetles account for roughly 25% of all animal species.

★ ★

Melanophila beetles can only lay their eggs in freshly burnt trees, so they are constantly seeking out fire. To locate fire, they use specialised sensors that can detect infrared radiation emitted by hot objects up to 130 kilometres away. These sensors are so sensitive that they're able to perceive changes in heat flux as minute as 0.13 milliwatts per square metre—a level of precision comparable to radio telescopes used in astronomy.

★ ★

During the 1940s, Melanophila beetles constantly swarmed a gridiron stadium in Berkeley, California, because they were drawn to the 20,000 patrons lighting cigarettes.

★ ★

You can determine the temperature by listening to the chirps of the snowy tree cricket. To calculate the temperature in degrees Celsius, count the number of chirps in an eight-second period and add five. For degrees Fahrenheit, count the number of chirps in a 15-second period and add 40. These formulas work because a cricket's metabolic rate increases proportionally with temperature, leading to faster muscle contractions and more frequent chirping.

★ ★

Depending on the species, periodical cicadas emerge from their underground habitat every 13 or 17 years—intervals that are both prime numbers. These unique surfacing schedules help ensure that different broods of cicadas rarely meet, overlapping just once every 221 years. Scientists believe these prime-number emergence cycles evolved to prevent the interbreeding between broods, thereby reducing the likelihood of hybridisation and preserving the distinct traits of each species.

★ ★

In 2024, the 13-year and 17-year cicada broods appeared above ground simultaneously in Indiana. The last time these broods emerged together was 221 years prior, in 1803. During the 2024 emergence, an estimated one trillion cicadas surfaced, with a combined weight of two million tonnes—equivalent to the weight of the entire human population of Australia.

★ ★

A single plague of cicadas can produce 500 trillion eggs. That's 1,000 times more than the number of stars in the Milky Way Galaxy.

★ ★

Arachnocampa luminosa is a glowworm native to New Zealand that lives on sheltered surfaces, such as cave ceilings and overhanging banks. It is famous for emitting a natural bioluminescent light that mimics the appearance of stars. When these glowworms are gathered together in a group, they create the stunning illusion of a starry night sky. This clever deception tricks their prey into believing they're outdoors, luring them to fly directly upwards—straight to the waiting glowworms.

★ ★

Contrary to popular belief, a chrysalis is not something the caterpillar builds or crawls into—it's actually the body of the caterpillar itself.

★ ★

During a caterpillar's transformation into a butterfly, it turns almost entirely into liquid. If you were to cut the chrysalis open at this time, a liquid caterpillar would ooze out. Despite undergoing this dramatic physical change, the butterfly that emerges from the chrysalis retains some of the experiences it learned as a caterpillar.

★ ★

Every year, four generations of monarch butterflies live and die during their annual migration across North America. When they arrive in central Mexico, many of them—despite having never been there before—miraculously settle on the very same tree that their great-great-grandparents landed on one year prior.

★ ★

Monarch butterflies fly southwards in swarms during their annual migration from southern Canada to central Mexico. However, for part of this journey, they take an unexpected detour eastwards over Lake Superior. This deviation makes their trip longer and more challenging because it takes them off their direct southwards path. Scientists believe the monarch butterflies take this detour because the area around this part of Lake Superior was once home to one of the tallest mountains in North America. Instead of expending energy flying over the mountain, the butterflies adapted by veering around it. Even though the mountain no longer exists, millions of generations of monarch butterflies have continued to follow this route.

★ ★

The butterflies living in the western Amazon rainforest are desperate for sodium—a scarce ingredient in the local ecosystem. To obtain this sodium, they regularly land on the heads of turtles to drink their tears. Surprisingly, the turtles remain still, letting the butterflies sip their salty liquid.

★ ★

The global insect mass is declining by 2.5% per year. If this trend continues, insect populations could completely collapse within a century, leading to catastrophic consequences for the ecosystem and human societies.

Parasites

A flea can jump 200 times its own body length in a single bound. That's equivalent to a human jumping over the Eiffel Tower.

★ ★

When a flea leaps into the air, it achieves a peak acceleration of 140 g. That's more than 40 times the force experienced by the astronauts during the Space Shuttle's launch and re-entry.

★ ★

Most aphids are born pregnant.

★ ★

Ear mites instinctively know not to infest both ears of a moth. Doing so would render the moth completely deaf, jeopardising the survival of both the moth and the mites themselves.

★ ★

If a person is bitten by a Lone Star tick, they can develop a lifelong allergy to red meat. This occurs because the tick injects a molecule called alpha-gal into their bloodstream. Their immune system then recognises alpha-gal as a threat and produces antibodies against it. Since alpha-gal is also found in red meat, the person experiences an allergic reaction whenever they consume red meat.

★ ★

Your face is likely teeming with thousands of tiny arachnids known as Demodex mites. These creatures, which have thin bodies and short legs, live within your pores and hair follicles. Measuring 0.2 to 0.4 millimetres in length—about one-third to one-fifth the width of a human hair—they are technically large enough to be seen with the naked eye. Yet, they remain invisible because they are nearly transparent, reside just beneath the skin's surface, and move extremely slowly, primarily at night. Nearly all adults have Demodex mites, and their numbers generally increase with age.

For most people, Demodex mites are harmless. However, high populations are associated with skin conditions such as blepharitis and acne rosacea.

★ ★

Toxoplasma gondii is a parasite that infects warm-blooded animals—including humans—and hijacks their brain to manipulate their behaviour. People are most likely to catch it through contact with cat faeces or by eating uncooked meat. It's estimated that up to half the world's population may be infected with Toxoplasma gondii. Studies have linked infection to an increased risk of suicide, mood disorders, road rage, and traffic accidents. In fact, some researchers estimate that Toxoplasma gondii could be responsible for at least one million deaths per year. That would make it as deadly as malaria.

★ ★

A 2024 study found that those who own a cat before they turn 25 are twice as likely to develop schizophrenia-related disorders. This increased risk is primarily attributed to Toxoplasma gondii infection.

Worms

In a 2013 study, researchers trained flatworms to navigate through rough terrain, then decapitated them. Remarkably, two weeks later—after their heads had regenerated—the flatworms still remembered how to traverse the rough terrain. It's not known how they were able to recall their training. Some researchers speculate that a flatworm's memories are stored somewhere in its body, while other researchers propose that the original brain of a flatworm modifies its nervous system in such a way that it retains information. The exact mechanism behind this phenomenon remains a mystery.

★ ★

A flatworm can be cut into 279 pieces, and each piece will regenerate into a completely new worm.

★ ★

The bootlace worm is the world's longest animal. Despite being just five millimetres wide, it can grow up to 55 metres in length.

★ ★

Eighty percent of all animals on Earth are nematodes. These tiny, worm-like creatures live in almost every environment, from the soil under our feet to the depths of the ocean. They play an important role in breaking down dead material and recycling nutrients.

★ ★

It's estimated that 4.4×10^{20} nematodes live in the top 20 centimetres of Earth's soil. That equates to between one million and ten million nematodes in every square metre of topsoil.

★ ★

What do you think is the average number of legs for an animal on Earth? Surprisingly, it's less than 0.1. That's because the vast majority of animals are nematodes, which are legless organisms.

Their sheer number is so overwhelming that they bring down the global average number of legs to nearly zero.

★ ★

In 2023, scientists successfully revived a nematode that had been frozen in the Siberian permafrost for 46,000 years. To put this timeframe into perspective, when the nematode was first frozen, Neanderthals still roamed the Earth.

Spiders

On 31 October 1832, Charles Darwin was sailing 100 kilometres off the coast of Argentina when he witnessed an extraordinary event: thousands of tiny red spiders began falling from the clear sky onto the HMS Beagle. More recently, spiders have been observed at altitudes of up to four kilometres above Earth's surface and as far as 1,600 kilometres from land. These incredible journeys were made possible due to a phenomenon known as ballooning. To balloon, a spider releases silk threads into the air, which catch the wind and act like a parachute. The threads are then lifted by air currents and Earth's electric field, enabling the spider to drift through the air.

★ ★

Spiders can walk on water. Their lightweight bodies and specially adapted legs distribute their weight evenly, so they don't break the water's surface tension.

★ ★

A spider's silk begins as liquid inside its body. This liquid is pushed through a small opening in the spider's abdomen, called a spinneret, which connects the protein molecules, transforming the liquid into a fine, threadlike silk.

★ ★

Pound for pound, a spider's silk is stronger than steel.

★ ★

Spider's silk is so strong that just a single thread—one kilometre long and 21 centimetres in diameter—could stop a Boeing 747 in mid-flight.

★ ★

Every year, spiders globally consume between 400 million and 800 million tonnes of prey. That's more than the total amount of meat and fish eaten by humans over the same period.

★ ★

In just one year, spiders could theoretically consume every human on Earth.

Rodents

A mouse can squeeze through a hole the width of a pencil.

★ ★

A 2014 study found that the scent of a male human is as stressful to mice as a three-minute swim, while the scent of a female human does not affect mice at all. This finding has important implications for the numerous studies that use mice as subjects. It suggests that the presence of male researchers could be a major factor behind the lack of reproducibility in animal research.

★ ★

Rats can outperform humans in logical decision-making. In an experiment, both rats and humans were tasked with predicting whether a red or green light would flash next in a series of trials. Each correct guess earned the participant a food reward. Although the lights flashed in a random sequence, the red light was programmed to flash 80% of the time, while the green light was programmed to flash just 20% of the time. The rats quickly learned to play the odds, consistently choosing the red light and achieving close to an 80% success rate. In contrast, the human participants attempted to identify patterns in the sequence, leading to a lower success rate of just 68%.

★ ★

A squirrel can survive a fall from any height, thanks to its low body mass and large surface area.

★ ★

Over the course of a single season, a squirrel can bury 3,000 nuts and successfully recall the locations of 90% to 95% of them.

★ ★

Prairie dogs utilise Bernoulli's principle to ensure a steady airflow through their burrow. When they construct their burrow, they build two entrances: one on a raised mound and the other on flat ground. This clever design ensures that when the wind hits the mound, it accelerates, creating a region of low pressure. In contrast, when the wind passes over the flat ground, it moves more slowly, creating a region of high pressure. The resulting pressure differential drives a one-way flow of air through their burrow, keeping it well-ventilated.

Dogs

Dogs possess such a strong sense of smell that they can detect human fingerprints—even after they've been left for three weeks.

★ ★

In ideal conditions, dogs can smell people and objects up to 20 kilometres away.

★ ★

Dogs can smell the difference between two similar fatty acids—caproic acid and caprylic acid—even when each is diluted to just one part per million. The only difference between these acids is that caprylic acid's main molecular chain is two carbon atoms longer.

★ ★

Dogs can smell when a person is stressed. They do this by detecting changes in specific hormones, like cortisol and adrenaline, which are secreted through a person's sweat and breath.

★ ★

Dogs can smell various medical conditions in humans, including lung cancer, breast cancer, hypoglycemia, narcolepsy, Parkinson's disease, malaria, and even COVID-19.

★ ★

Dogs can smell an epileptic seizure in a human up to 68 minutes before it occurs.

★ ★

When a dog encounters a scent, it tends to initially sniff it using just its right nostril. This is because the right nostril is connected to the brain's right hemisphere, which processes new experiences and potential threats. If the scent turns out to be familiar or non-threatening, the dog will switch to using its left nostril.

★ ★

In a 2018 study, researchers placed a bar magnet inside one of three opaque jars. Dogs were able to identify the jar containing the magnet at a rate significantly higher than chance.

★ ★

Dogs find it difficult to watch cathode-ray tube television. A cathode-ray tube television has a maximum refresh rate of 60 hertz, meaning it displays 60 images per second. This is suitable for the human eye because it perceives just 30 to 60 images per second.

However, it's not suitable for the dog eye, which perceives 75 images per second. When a dog watches a cathode-ray tube television, it sees gaps between frames, which appear like an annoying flickering light. This makes the viewing experience unpleasant. Dogs prefer to watch modern televisions because they have a refresh rate of around 120 hertz, meaning there is no flickering effect and the motion appears smooth and natural.

★ ★

In a 2015 study, researchers explored how dogs perceive human social behaviour. Dog owners were asked to pretend that they were struggling to open a jar. An actor then interacted with the owner in one of three ways: helping with the jar, watching on, or refusing to assist. Following the interaction—which the dogs had been observing—the actor offered food to the dogs. The results showed that the dogs were significantly more likely to accept food from the actor who helped the owner, or even the actor who just watched on, while they tended to reject food from the actor who refused to assist. This study indicates that dogs make social judgements about people based on their actions.

★ ★

Research shows that when Earth's magnetic field is calm, dogs tend to align themselves with the north-south axis when they defecate.

★ ★

Dog urine contains compounds—including urea, salts, and ammonia—that are highly corrosive to metals over time. As a result, when a dog repeatedly urinates on a metal lamppost, it can collapse.

★ ★

Scientists have created an anti-aging drug for dogs called LOY-002, which targets age-related metabolic dysfunction. Clinical trials of LOY-002 are underway, and if successful, the drug is expected to be commercially available within the next few years. A second team of scientists is trialling another anti-aging drug for dogs called rapamycin, which has already been shown to have life-extending effects in rodents, worms, and flies. It works by blocking a molecule called mTOR, which regulates cell proliferation and death. It's believed that the science behind life extension in dogs may also be relevant to human health.

★ ★

Slightly underweight dogs live an average of two years longer than overweight dogs.

Cats

Just one out of every 3,000 tortoiseshell cats is male. This gender imbalance arises because the tortoiseshell's distinctive orange and black coat requires two X chromosomes to develop. Since males typically carry only one X chromosome (XY), it's extremely rare for them to exhibit the tortoiseshell pattern.

★ ★

It's relatively common for kittens in the same litter to have different fathers. This phenomenon, known as superfecundation, occurs because female cats in heat often mate with multiple males in a short period. Since the female's eggs can be fertilised at different times, kittens from the same litter often end up with different fathers.

★ ★

In 1970, researchers conducted an experiment to explore how early visual experiences shape a kitten's perception. They raised one group of kittens in an environment filled exclusively with horizontal lines and another group of kittens in an environment filled exclusively with vertical lines. When the kittens were later introduced to a chair, those raised with horizontal lines could see the seat and jump onto it, but they repeatedly bumped into the chair's legs. Contrastingly, the kittens raised with vertical lines could navigate around the chair's legs but struggled to perceive and jump onto the seat. This study provides evidence that the brain undergoes a critical window of development, during which it must be exposed to certain sights, otherwise it may remain blind to them forever.

★ ★

In a 1954 experiment, researchers removed cats from their homes, transported them in sacks, and placed them in the centre of a circular, covered maze that had exits spaced every 15 degrees. Remarkably, most of the cats chose to exit the maze in the direction of their home. The cats were even more likely to exit the maze in the direction of their home if they lived fewer than five kilometres away. These findings suggest that cats have an internal homing mechanism that doesn't rely on visual cues.

★ ★

During the 1960s, neuroscientist Michel Jouvet and his team conducted a range of experiments on cats in which they lesioned the specific brainstem regions that induce muscle paralysis during REM sleep. This prevented muscle paralysis from occurring, so the cats acted out their dreams. The cats were observed stalking, running, pouncing, and clawing while asleep. The results of this experiment suggest that cats dream about hunting-like activities.

★ ★

In the early 1980s, the average lifespan of a pet cat was only seven years. Today, that number has doubled to approximately 14 years. This significant increase in longevity is primarily due to improvements in the nutritional quality of cat food and advancements in veterinary care.

Elephants

Elephants can detect low-frequency vibrations through their feet and trunk—vibrations that are imperceptible to humans. This allows them to sense when a distant storm is approaching. Interestingly, if an infrasonic speaker is played to an elephant, it will turn its back to the speaker, thinking it's a storm.

★ ★

When a mother elephant senses that a predator is nearby, she protects her baby by stomping her feet. This action generates seismic waves that can travel up to 30 kilometres across the Earth's surface. Distant elephants can detect these subtle vibrations through their feet and trunk. This form of communication allows elephant herds to stay alert to potential dangers, even when they are far apart.

★ ★

An elephant can smell a water source from up to 19 kilometres away.

★ ★

In a 2019 study, researchers presented elephants with two buckets containing different quantities of sunflower seeds. Relying solely on their sense of smell, the elephants consistently selected the bucket with more seeds. Some elephants could even reliably distinguish between one bucket containing 150 seeds and another containing 180 seeds—a difference so subtle that the human eye might struggle to distinguish it.

★ ★

In a 2014 study, researchers investigated whether elephants could differentiate between human languages. They played voice recordings to the elephants: one in the language of the Maasai men, who occasionally hunt elephants, and another in the language of the Kamba men, who don't harm elephants. The elephants reacted significantly more defensively to the Maasai language than to the Kamba language. This suggests that elephants are capable of distinguishing between human languages.

★ ★

Elephants are excellent distance swimmers. They have been recorded swimming nonstop for six hours, covering distances of up to 48 kilometres.

★ ★

Elephants are the only wild animal that can understand human pointing without prior training.

Primates

More than 30 million years ago, monkeys made an incredible journey from Africa to South America. Scientists believe this migration took place when African monkeys living near the coast were accidentally swept out to sea on a natural raft composed of vegetation and debris. Over several weeks, ocean currents carried this raft across the Atlantic Ocean, eventually depositing it 1,500 kilometres away in South America. While this event had an extremely low probability of happening, over vast stretches of time, even rare events do take place.

★ ★

Japanese macaque monkeys have been observed riding deer for both fun and as a means of transportation. In return, the monkeys feed and groom the deer, creating a mutually beneficial relationship between the two species.

★ ★

The bonobo is the only non-human animal that regularly copulates face-to-face.

★ ★

In a 2014 study, researchers compared the strategic abilities of chimpanzees and humans using a game similar to *Rock, Paper, Scissors*, where the participants had to adapt their strategy based on their opponent's moves. Surprisingly, the chimpanzees worked out the optimal strategy significantly faster than the humans.

★ ★

In a 2007 experiment, chimpanzees and humans played a game in which the numbers from 1 to 9 appeared on a screen in random locations before disappearing and being replaced by blank boxes. The participants had to recall the positions of the numbers by touching the boxes in ascending order, from 1 to 9. Surprisingly, young chimpanzees outperformed university students at this game.

★ ★

In a 2011 study, researchers challenged chimpanzees and human children to retrieve a food reward from the bottom of a fixed, transparent tube. Twenty percent of the chimpanzees figured out that they could use water from a nearby drinking fountain to solve the problem. They filled their mouth with water, spat into the tube, and floated the reward to the top to retrieve it. In contrast, only 8% of the four-year-old human children—who had been given a jug to assist them—successfully retrieved the reward.

★ ★

In 1931, comparative psychologist Winthrop Kellogg conducted an experiment to determine whether a chimpanzee could learn human behaviours. Kellogg raised a seven-month-old chimpanzee named Gua alongside his ten-month-old son, Donald. Surprisingly, Gua initially outpaced Donald in learning human-like behaviours. She quickly mastered skills like drinking from a glass, eating with a spoon, opening doors, skipping, and understanding the meaning of several words—milestones that Donald had not yet achieved. However, by the time both Gua and Donald turned one, Donald's development accelerated rapidly, and he began to surpass Gua in communication and language comprehension. The experiment ended after nine months, when Donald started making chimpanzee noises.

★ ★

In 2010, a chimpanzee named Julie inserted a long-stemmed blade of grass into her ear and left it there. Although the grass served no practical purpose, the other chimpanzees in Julie's group copied her. This behaviour continued, even after Julie's death. Primatologists believe that by wearing a blade of grass, Julie was making a fashion statement, similar to how humans wear earrings or a necklace. This behaviour spread through the group like a cultural trend, mirroring the way fads catch on in human societies.

★ ★

Some scientists believe it's possible to create a human-chimpanzee hybrid, known as a humanzee. During the 20th century, a Soviet biologist and a team of Chinese researchers independently attempted to produce such a creature, but neither group succeeded.

Extinct Animals

An estimated 99.99% of all the species that have ever existed are now extinct.

★ ★

The largest flying animal to ever exist was a pterosaur called Quetzalcoatlus northropi. It was as tall as a giraffe and had a wingspan of 11 metres—wider than an F-16 jet. Despite its enormous size, it was able to fly because it had hollow bones, specialised wing membranes, powerful flight muscles, and an efficient respiratory system.

★ ★

Brazil is home to thousands of massive underground tunnels, the largest of which measures 600 metres long, 1.8 metres tall, and 1.5 metres wide. It's believed that these tunnels were dug as recently as 10,000 years ago by giant ground sloths. These enormous creatures, which were as large as modern African elephants, used their powerful claws to carve out the underground passages. The claw marks from this activity are still visible on the tunnel walls today.

★ ★

Woolly mammoths still inhabited the Earth 1,000 years after the Great Pyramid at Giza had been constructed.

★ ★

Palaeoloxodon falconeri was a species of dwarf elephant that grew to just one metre in height. It inhabited several Mediterranean islands, where its small size likely evolved due to limited space and resources—a phenomenon known as insular dwarfism.

★ ★

Whales weren't always the water-dwelling mammals that we know today. Around 50 million years ago, the first whales had webbed feet and lived on land.

★ ★

The megalodon was an ancient mega-shark that grew up to 20.3 metres in length—longer than a standard bowling lane. Its enormous teeth, measuring up to 17.8 centimetres, were once believed to be petrified dragon tongues.

★ ★

Titanoboa cerrejonensis was a massive prehistoric snake that lived 60 million years ago in the swamps of Colombia. It grew up to 14.3 metres in length—longer than a school bus. The thickest part of its body was 90 centimetres in diameter—roughly the size of a truck's tyre.

★ ★

Stupendemys geographicus was a giant freshwater turtle that once inhabited the lakes and rivers of South America. It grew up to four metres long and weighed as much as 1,100 kilograms—comparable to the size and weight of a car.

★ ★

Palaeeudyptes klekowskii was the largest penguin species that ever existed. It grew up to two metres tall and weighed as much as 116 kilograms, making it roughly the same height and weight as basketball player LeBron James.

★ ★

During the Carboniferous period, Earth's atmosphere contained roughly 10% more oxygen than it does today. This increased oxygen level caused ancient insects to grow much larger than their modern counterparts because it enabled more efficient respiration through their tracheal system.

★ ★

The dragonfly-like creature Meganeuropsis permiana was the largest insect that ever existed. It had a wingspan of about 70 centimetres, making it roughly the same size as a modern-day pigeon.

★ ★

Arthropleura was a giant millipede-like arthropod that lived during the Carboniferous period. It holds the title as the world's largest known land invertebrate, growing up to 2.6 metres in length.

★ ★

We will never know anything about the vast majority of extinct animals. This is because fossilisation is such a rare process that an estimated one bone out of every billion becomes fossilised. To put this figure into perspective, the entire US population alive today will leave behind a total of just 70 bones—about one-third of a human skeleton. Consequently, it's estimated that less than 0.1% of all the animal species to have ever existed have become fossils. Of these, only a small fraction will ever be discovered.

Dinosaurs

Dinosaurs from the early Cretaceous period roamed the Earth when it was situated on the opposite side of the Milky Way Galaxy.

★ ★

The earliest dinosaurs emerged 100 million years before flowering plants existed.

★ ★

Stegosaurus lived between 155 and 150 million years ago, while Tyrannosaurus rex appeared much later, existing from 73 to 66 million years ago. This means that Tyrannosaurus rex lived closer in time to humans than it did to Stegosaurus.

★ ★

It's estimated that a total of 2.5 billion Tyrannosaurus rex have existed throughout history. Despite this immense population, only about 20,000 adult Tyrannosaurus rex were alive at any one time, across roughly 127,000 generations.

★ ★

If the Tyrannosaurus rex population had been 2.5 million, instead of 2.5 billion, we probably wouldn't know they existed. This is because a smaller population would have made it extremely unlikely for us to ever discover a Tyrannosaurus rex fossil.

★ ★

Tyrannosaurus rex is more closely related to modern hummingbirds than it is to Stegosaurus.

★ ★

Bruhathkayosaurus was a sauropod that might have been the largest land animal ever. It grew up to 45 metres in length, making it almost as long as an Olympic-sized swimming pool.

★ ★

Sauropods had individual sensory nerve cells that extended all the way from their brainstem to the tip of their tail. These are believed to be the longest cells in any animal throughout history, measuring up to 45 metres in length. Due to the immense size of these cells, it took half a second for a signal to travel from the tip of a sauropod's tail to its brain.

★ ★

Some species of pterosaur had heads that were four times the length of their body.

★ ★

Dinosaurs didn't roar. They likely made cooing sounds, similar to modern-day birds. The idea of roaring dinosaurs is largely a Hollywood invention.

★ ★

Non-avian dinosaurs inhabited the Earth from 245 million to 66 million years ago. This means they existed for a longer time than they have been extinct.

★ ★

No dinosaurs lived in the ocean.

★ ★

Pterodactyls were not dinosaurs. They were actually a group of flying reptiles that lived during the same period as the dinosaurs.

★ ★

A new species of dinosaur is discovered every two weeks.

★ ★

The vast majority of dinosaur skeletons in museums are not the original fossils—they are just artificial casts.

Mammalian Traits

Every mammal weighing over three kilograms takes about 21 seconds to urinate. The reason for this consistency lies in the design of the mammalian urinary system—it can be scaled up by a factor of 3,600 in bladder volume without compromising its function.

★ ★

Nearly all mammals experience one billion heartbeats in their lifetime. Smaller mammals, like mice and shrews, have faster heart rates and shorter lifespans, while larger mammals, like elephants and whales, have slower heart rates and longer lifespans. Humans are a unique exception to this rule—our lifespan is significantly longer than what would be expected based on our heart rate.

★ ★

Mammals typically have twice as many teats as their average litter size. For example, humans have two breasts and an average litter size of one, while cats usually have eight teats and an average litter size of four.

★ ★

All mammals, with the exception of sloths and manatees, have exactly seven vertebrae in their neck. This means that although a mouse and a giraffe have vastly different neck lengths, they still share the same number of neck vertebrae.

★ ★

In 2021, a team of Japanese scientists discovered that some mammals can breathe through their anus. In their experiment, they deprived mice, rats, and pigs of oxygen and then administered them an oxygen-rich enema. The blood vessels in the animals' rectal lining absorbed the oxygen and circulated it through their bloodstream, keeping them alive.

★ ★

The blue whale, the world's heaviest mammal, weighs 100 million times more than the Etruscan shrew, the world's lightest mammal.

★ ★

Bats are the only mammals capable of flight. However, since bats account for 20% of all mammal species, this means that 20% of all mammal species can fly.

General Animal Traits

Peto's paradox is the observation that larger species are not more prone to cancer than smaller species. It is a paradox because larger species are composed of a greater number of cells, so they should have a higher incidence of cancerous mutations. As an example of Peto's paradox, whales have roughly 1,000 times as many cells as humans, yet they actually experience a lower rate of cancer.

★ ★

Smaller animals tend to experience time more slowly. This is evidenced by the fly. It perceives 250 frames per second—about four times as many as humans—so it views the world four times as slowly. This heightened perception allows it to detect and respond to threats incredibly fast, making it notoriously difficult to swat.

★ ★

In general, the smaller the species, the longer its sperm. An example of this phenomenon can be seen in the tiny fruit fly, Drosophila bifurca, which produces the longest sperm in the animal kingdom. When uncoiled, its sperm measures 5.8 centimetres—more than 20 times the length of its body.

★ ★

The longer an animal yawns, the larger its brain tends to be. This correlation may be explained by the brain-cooling hypothesis, which proposes that yawning helps reduce the brain's temperature by increasing blood flow and drawing in cooler air. Since animals with larger brains produce more heat due to higher neural activity, they require longer yawns to effectively cool their brain.

★ ★

Gloger's rule states that warm-blooded animals living in hot, humid regions tend to be darker in colour than those living in cold, dry regions. An example of this rule can be observed in house sparrows—those from tropical regions have darker plumage than those from northern Europe.

★ ★

Bergmann's rule states that, within a taxonomic group, larger animals tend to inhabit colder regions, while smaller animals tend to inhabit warmer regions. A clear example of this rule can be seen in penguins. The emperor penguin, the largest of all penguins, lives in the chilly environment of Antarctica, whereas the Galapagos penguin, one of the smallest penguin species, lives in the warm climate of the Galapagos Islands.

★ ★

A 2020 study found that farm animals begin to change their behaviour up to 20 hours before an earthquake. Some scientists propose that this is due to the animals sensing the electromagnetic fluctuations in the atmosphere that often precede seismic activity.

★ ★

Parthenogenesis is a type of asexual reproduction where a female can produce offspring without fertilisation from a male. Normally, a female's egg requires genetic material from a male to initiate development, but in parthenogenesis, the egg uses its own genetic material to start the process. As a result, the offspring produced in parthenogenesis are genetically very similar, or even identical, to the mother. Parthenogenesis occurs naturally in some insects, fish, amphibians, birds, and reptiles. Notably, the Komodo dragon, which can grow up to three metres in length, is the largest known vertebrate capable of reproducing through parthenogenesis.

★ ★

Hibernation and sleep are not the same thing. During hibernation, an animal's metabolic activity drops significantly, but its brain does not enter a true sleep state. As a result, when an animal emerges from hibernation, it is typically sleep-deprived and needs to catch up on proper sleep to fully recover.

★ ★

Flight has evolved separately at least four times throughout history: in insects, pterosaurs, birds, and bats.

★ ★

Bioluminescence—the ability of a living organism to produce and emit light through chemical reactions—has evolved separately at least 94 times in the animal kingdom.

Miscellaneous Zoology

The Paratarsotomus macropalpis mite is the fastest animal in the world relative to its size. It can move at 322 body lengths per second—the equivalent of a human running at Mach 1.7.

★ ★

Tardigrades are eight-legged micro-animals known for their extraordinary resilience. They can withstand high levels of radiation, pressures six times greater than those found in the deepest ocean trenches, and even being shot out of a gun. Additionally, they are the only known multicellular animal that can survive in the vacuum of outer space. When faced with extreme conditions, they enter a death-like state known as cryptobiosis. To reach this state, they expel more than 95% of the water from their body, retract their head and legs, curl into a tight ball, and drastically reduce their metabolic activity to just 0.01% of normal levels. They can remain in this cryptobiotic state for more than 30 years.

★ ★

The wombat is the only animal that produces cube-shaped faeces. This unique shape is less likely to roll away, allowing wombats to effectively mark their territory and communicate with other wombats through their faecal scent.

★ ★

Platypuses are extraordinary creatures: they possess ten sex chromosomes, glow under ultraviolet light, and hunt by detecting the subtle electric fields produced by their prey.

★ ★

Mother platypuses secrete milk through the pores in their skin. The milk then collects on their fur, where their babies lap it up directly.

★ ★

Sea otters have a favourite rock, which they store in their pouch. They use this rock as a tool to crack open molluscs and clams for food.

★ ★

When a tiger salamander population becomes too high, some offspring develop larger bodies, wider mouths, and bigger teeth. These physical changes, known as a cannibalistic morph, enable them to eat other salamanders, which helps regulate the population.

★ ★

Atrazine is a widely used herbicide that can transform male frogs into females. It severely disrupts the endocrine system, thereby causing male frogs to develop female characteristics, including the ability to reproduce as females.

★ ★

A polar bear can smell a seal sitting on ice up to 32 kilometres away.

★ ★

A grizzly bear's bite is so powerful that it can crush a bowling ball.

★ ★

A pizzly bear is a hybrid between a polar bear and a grizzly bear. Over the past decade, these unique creatures have been appearing more frequently due to the impacts of climate change. As the Arctic sea ice shrinks, polar bears are being forced to venture south in search of food. This migration is bringing them into contact with grizzly bears. As polar bears and grizzly bears only diverged 500,000 years ago, they are still able to mate and produce viable pizzly bear offspring.

★ ★

Sloths can hold their breath longer than dolphins.

★ ★

Sloths have the slowest digestion of any mammal, taking up to one month to process a single leaf. As a result, they only need to defecate once a week. During this process, they climb down from their tree and relieve themselves on the ground. However, descending from a tree is a risky endeavour for a sloth; in fact, 50% of all sloth deaths in the wild occur during this weekly trip to the ground. When a sloth does defecate, it can lose up to one-third of its entire body weight.

★ ★

A 2008 study found that when cows are grazing or resting, they typically orientate their body in a north-south direction to align with Earth's magnetic field.

★ ★

A cow's rumen—its largest stomach compartment—can hold up to 190 litres of food and water. That's equivalent to the volume of an average bathtub.

★ ★

Cows, sheep, and goats don't have upper front teeth. Instead, they have a dental pad made of hard gum tissue.

★ ★

Mountain goats can climb slopes as steep as 70 degrees.

★ ★

The Pyrenean ibex is a type of wild goat that is the first extinct species to be brought back to life—and the first to go extinct twice. The world's last Pyrenean ibex died in 2000, when it was crushed by a falling tree. In 2003, scientists successfully cloned the Pyrenean ibex, briefly reviving the species. This newborn Pyrenean ibex lived for only several minutes before it died due to a lung defect, making the species extinct for a second time.

★ ★

Female African buffaloes engage in a type of voting behaviour. When the herd is resting, individual females will stand up and face the direction they believe the herd should move. Once the females have cast their vote, the herd treks in the direction favoured by the majority.

★ ★

Deer antlers are the fastest-growing tissue of any mammal, growing up to 2.75 centimetres per day.

★ ★

Giraffes have a unique courtship process. The male begins the proceedings by rubbing the female's hindquarters to encourage her to urinate. He then leans down to drink her urine, and if he tastes a hormone that indicates she's ovulating, he proceeds with the mating ritual.